YOUTH:
change
and
challenge

YOUTH:
change
and
challenge

edited by ERIK H. ERIKSON

BASIC BOOKS, Inc., Publishers

NEW YORK   LONDON

Fifth Printing

Library of Congress Catalog Card Number 62-17242
Manufactured in the United States of America

Most of the essays in this book originated in a seminar sponsored by the Tamiment Institute and held in Tamiment-in-the-Poconos in May 1961. They were first published in Dædalus, the journal of the American Academy of Arts and Sciences.

CONTENTS

Editor's Preface

In the winter of 1961-1962, *Dædalus*, the journal of the American Academy of Arts and Sciences, brought out an issue on youth (here to be referred to as "the issue"). It forms the basis for and provides the bulk of this book (here to be called "the book"). And although the privilege of writing the preface to the book has been given to me, it should be clear that the editor of *Dædalus*, Professor Stephen Graubard, and his co-workers planned and executed the issue in that special manner which has become a *Dædalus* tradition and works toward a high combination of scholarly endeavor and personal style. Having been selected at a preliminary conference, the potential contributors to an issue of *Dædalus* are invited to foregather for a number of days in quiet surroundings in order to confront each other and their first drafts. The publications which emerge from such gathering of minds are modern symposia and should be read as symposia, the strength—and the limitations—of which are integral to the composition of the group thus collaborating.

The issue on youth was sponsored by the Tamiment Institute and its Educational Director, Mr. Norman Jacobs. The conference was held in the Pennsylvania hills, at Tamiment-in-the-Poconos. Mr. Ben Josephson was the host and Professor Daniel Bell served as chairman.

"The questions treated in this issue," so Stephen Graubard concludes his preface to the issue, "are too complex to admit of any definitive answer"; and it may fairly be said that every special issue of *Dædalus* is a deliberate inauguration of a discussion long to be continued by the contributors, the readers, and the critics.

The role of secondary editor provides an advantage not shared by primary ones. It permits a more leisurely retrospection on the origin and the development of the symposium; and it adds to such perspective the intelligence (reported by informants around the bookstands of Harvard Square) that the issue of *Dædalus* was a popular one among young as well as older readers. I know, in fact, of a college which has handed the issue to its freshman class. I would like to say a few words, then, on the special character of a book written by one section of humanity about another—and read by the other.

Youth shares with other groups, such as women and old people, the fact that the role assigned to it by nature has been elaborated by cultures as a set of differences from some standard human being, the norm, of course, being usually the normal adult male. The group is then judged on the basis of what it is not and will never be, or is not quite yet, or is not any more. The assignment of such an arbitrary measure of worth results in an intensification of the question of free will, of possible choices within a position thus prejudged, and of the chance of ever being uniquely oneself. The very intensity of these questions can make such a group alternately over-eager to be discussed and over-sensitive to being discussed as different; and it can aggravate the very defensiveness attributed to the group by its observers and judges.

It seems important, therefore, to specify in each case who is writing what about whom and using what measure. If in this pursuit we become aware of the fact that more than half of the contributors to this book have lived for more than half a century, we do so not in order to get set for an argument *ad hominem* but for a general orientation in this matter of one generation writing about another. Today, men over fifty owe their identity as individuals, as citizens, and as professional workers to a period when change had a different quality and when the dominant view of the world was one of a one-way extension into a future of prosperity, progress, and reason. If they rebelled, they did so against details of this firm trend and often only for the sake of what they thought were even firmer ones. They learned to respond to the periodical challenge of war and revolution by reasserting the interrupted trend toward normalcy. What has changed in the meantime is, above all, the character of change itself, as will become obvious in this book. Is it, then, surprising that some of our contributions reflect the challenge which the new kind of change confronts *us* with; that they reflect, on occasion, a yearning for *our* kind of preoccupation with youthful trouble (note the nostalgic references to Hamlet) and reflect also a certain alienation from such sectors of modern youth (called alienated by us) who, far from wishing to play the role which we would recognize as youth, pass us by as unhelpful and, worse, irrelevant? Conversely, some types of youth chosen by us for special and warm attention appear to be those that are somehow like us or profess to need us, those who are responsive to us or for whom we feel responsible. They include youth at home eager to study and youth abroad ready to rebel, youth at home that feels harrassed by change and youth everywhere that withdraws loudly or rebels noisily. Verbal youth, then, claims our sympathy, as does youth standing out by idiosyncratic action.

Its specific burden of responsibility determines the image an older generation has of the younger one. A peculiar guilt can haunt those of the older generation—the guilt over having caused what they cannot guide to a foreseen completion. This can make them look at youth as a cast of characters looking for a scenario not yet written; or worse, a cast populating a play with a scenario already in production and badly in need of rewriting. Of all periods in history, our period of ruthless manufacture—led by men who called themselves self-made and felt called upon to overtake evolution itself with human reason—our energetic and yet sensitive period is overcast with guilt over what it has wrought. Much of youth, however, grew up in and with these changes and not only takes the new kind of change for granted but experiences technological advance as a "natural" challenge; it is quite willing to improvise the continuation of the scenario. It is, then, sometimes not too clear who is suffering most under the changes of our time, and it is lucky, indeed, that the contributors to this book refuse to dissimilate their own bewilderment as they face the task of stating where patterns of modern youth fit discernibly into their established conception of the world, be it based on the knowledge of evolution, history, or society. It is in such work of competent synthesis that the respect of the generations for each other comes through and forms a solid bond of transition.

In his preface to the issue, Stephen Graubard mentioned my role in the original planning, a role which explains this preface although I was not able to attend the conference. What I will do, then, is to compare our intentions (our myth, if you wish) with the reality of the issue. The general direction we agreed upon at the beginning is obviously encompassed by the words Change and Challenge. We remembered the selective emphasis given in today's academic, journalistic, and clinical publications to the problems caused for youth by rapid change; we reviewed the selection of deviant, exotic, and extremist aspects of youth in our magazines; we discussed the assumed alienation of the youth of our age from the inner and outer sources of true rejuvenation and spiritual survival; and we concluded that a symposium on youth should not only concern itself with the sources of such dire developments and their alleged universality but should also attempt to pay attention to that quiet, determined, and competent youth which seems to enter the ranks of leadership in big and small countries, in industrialized and as yet underdeveloped parts of the world. Is not change the business of youth and is not challenge the essence of its business? To understand youth and its position in the center of contemporary actuality, one should set one's sight on the advance outposts of its anticipated mastery before attempting to assess the nature of the dangers to which it is exposed and the regressions to which it succumbs.

If an editor may impose on his readers remarks slanted toward his own predicament, I should like to point to the fact that the more morbid trends in the discussion of youth are often said to result from the popularization and influence of clinical writings. And, indeed, the prevailing conception of youth in particular and of human motivation in general has been much influenced by a medical and largely psychiatric transvaluation of human phenomena. Maybe the simplest way to characterize this view is to refer to the typical medical report, in which the frequent occurrence of the word "negative" denotes the fact that the patient enjoys positive good health. The degree of positiveness and the range of vitality remain unspecified: enough that the examination was negative regarding a series of expectable defects and symptoms which one has learned to diagnose and to treat. (Quite parenthetically, we have a similar transvaluation in questions of loyalty: when going abroad for international service, we can expect, at best, to have our loyalty certified as not, at the moment, subject to "any reasonable doubt.") But I think that we clinicians would be either flattering or blaming ourselves unduly if we did not realize that much of our so-called negative influence is due to an unpredictable exploitation of our insights by prevailing cultural trends. That Freud should ever become the hero of a paperback culture—that he should be read as the prophet of a cultural mood rather than as the architect of painstaking thought to be built on painstakingly by future generations—this, I think, the Professor feared but could neither fully foresee nor attempt to forestall. We, his followers, however, reach from his world into ours, and while we may work in his shadow, we cannot claim that we work unseen; and while we continue to work in what he compared to deep mining shafts, we cannot ignore what happens to what we unearth and see on sale in a crowded and affluent market. We cannot, for example, write about youth without writing for youth. This may be good all around. If children could have looked over our shoulders, some of the more flamboyant interpretations of childhood might have remained unwritten.

To characterize our influence, however, as positive (and inspiring) or negative (and conducive to fatalism), does not do justice to the situation. There are matters of lasting concern which can be studied only by the psychoanalytic procedure and methodology and must be so studied even if our conclusions may lead us to assumptions at variance with the dominant aims and methods of our civilization. To abandon this obligation would render our whole work irrelevant. If we were to become the technicians of a general adjustment, we could not be of help to the very individuals for whom our method can do the most and we would cease to be the critics of the rationalizations and repressions of changing civilizations. On the other hand, one cannot study the unconscious except from the vantage point of contemporary historical actuality. Otherwise, the unconscious becomes a tableau for the romantic yearnings of a decaying period.

That some of us, then, with seemingly belated haste, turn to the "positive" side of human motivation and to an interest in contemporary actuality is not primarily an apologetic promise to mend our ways and to become strenuously inspirational. Rather, our field has developed to the point where the question of the relation of inner strength to situations of challenge, commitment, and decision calls for our concerted attention. For even as we must refuse to be technicians of adjustment, we should abjure being the harbingers of maladaptation. One cannot cure if one knows only disease; one cannot do research if one accepts deviations as the measure of all things. Nor can one teach if one believes only in that which is unteachable.

But there is an "analytic" trend discernible in our time (and in this book) which transcends the clinical emphasis just mentioned. Self-scrutiny has, in fact, become a world-wide trend, whether the official motivation and the preferred method is more one of depth psychology or of political "thought reform," of autobiographic self-examination or of existential meditation. Self-scrutiny, in turn, is aggravated by and contributes to role consciousness, and it is no coincidence that many of the historical changes which converge on youth as a pivotal generation concern the self-conscious creation of new roles and new styles within new technologies, new nationalisms, and new ideologies. From such a wider perspective we miss in the issue of *Dædalus* more information on the youth both at home and in foreign countries which increasingly superimposes universal aspirations and expectations on the most disparate national, religious, and cultural origins. The question as to how such youth reconciles its waning moralities and its pervading beliefs with its transient ideologies and the immediate ethical demands of a new productivity—all of this is approached but not fully treated as yet in this book.

The difficulty, then, which some of us have with the evaluation of the new technical elite (and rank and file) has a counterpart in the ease with which we understand other elites. One could almost speak of a conspiracy of angry youth and angry middle age, often instituted under the guise of radical modernism by disappointed ideologues. In contrast to youth cultures, youth movements depend on adult sanction and on the support of selected adults who aspire to be (or to lead) its leaders. As a college teacher I have been aware of such a "conspirational" trend, for example, in the embarrassing similarity between the *overt* behavior of some young individuals and the *hidden* motives ascribed to the genre "youth" by popular writers. If a teacher, to take an example, has written about "identity-diffusion" as a characteristic of the *dynamics* of the state of youth, he is apt to see the implicit theme become a chosen, explicit, and exaggerated role. This is, of course, part of the process of identity formation in a time which is so acutely identity *and* publicity conscious. As *West Side Story* has it: "We're cruddy juvenile delinquents, So that's what we'll give 'em."

What these "delinquents" sing and dance, however, the more introverted may live out in a more intellectual underground. Every "creative" subculture, of course, calls its own a number of adult men and women with identities that remain exposed to the fluctuation between the brooding isolation and the bursting communication that is characteristic of the intellectual and the artist. Here, youth and middle age feel congenial to each other as one struggles to conquer the means of creativity and the other to reconquer it. But whatever empathy and sympathy may unite youth with those who teach and care for them and think and write about them, the decisive factor is that the young ones should become competent in transmitting to *their* peers (strange as they may find them) what they themselves will decide to represent in a decisive future—whatever combination, that is, of leadership, of participation, and of creative isolation.

The sequence of articles in this book is the same as that in the issue of *Daedalus*. However, three articles have been added. Scoops are rare in academic life but we may consider it a bit of one that Justice Goldberg consented to write on legislation relevant to labor youth just before he resigned as Secretary of Labor. Two new contributors record their impressions of American youth, in the Peace Corps and in the Sit-In Movement—small groups who have chosen positions of extreme hardship under the eyes of the nation and of the world. Both contributors, Joseph Kauffman and Robert Coles, write with a participant heart, and the reader will not think the less of what they offer.

All in all, the contributions are roughly grouped in three sections. There are, first, a number of panoramic papers, which look at youth as an evolutionary and a generational, a societal and a national, phenomenon. Maybe the global approach of the contributors to the first section has been enhanced by the distance between their birthplaces and their places of work: Professors Eisenstadt, Naegele, and Bettelheim were born, respectively, in Warsaw, Stuttgart, and Vienna and are working in Israel, British Columbia, and Illinois. The papers of Eisenstadt and Parsons constitute a whole condensed course in sociological thought. Parsons' paper, at the same time, introduces the second group of contributors (all born, for simplicity's sake, in this country), who concern themselves primarily with American youth. Finally, there is a group of papers on the youth of other lands, namely, Japan, France, and Russia, all written by men who have lived and studied in these countries. Here, it is Professor Wylie who focuses his comparison back on American youth.

If I had to write a conclusion rather than a preface, as I fear I have already largely done in my role as secondary editor, I would point to a convergence in the papers of Justice Goldberg and Mr. Sherman—a convergence which may well be responsible for much of this book's intellectual critique of our time. For Justice Goldberg warns that even in our individualistic land it will be, in view of the emphasis on specialized labor, "difficult to maintain an exclusive preoccupation with individual problems," while Mr. Sherman wonders what will happen to the "revolutionary momentum" in Russia, where "calculation rather than spontaneity is a young person's guide to success." Yet Mr. Sherman also reports that the young poet Yevtushenko prefers to read his poetry to young scientists and engineers because of their "fresh minds."

Having read this extraordinary symposium the reader may yet be left with the impression (or else I intend to leave him with it) that there are two great themes which—although vigorously introduced by Talcott Parsons and Bruno Bettelheim—remain undeveloped: these are the themes of technological youth and of women. Alienation is part of the human condition, even if each age is rightly preoccupied with its own version. I would think that an age is not characterized by a greater or lesser degree of alienation, but by its own kind of tension between alienation and worldly initiative. A truly universal section of contemporary youth probably feels as at home in (or should one rather say "in the swing of") the challenges of our age as any youth, as far as we know, ever did in theirs. The second undeveloped theme is that of female youth. There appear to be themes of such strangeness that rational man will ignore them until a sudden startle tells him that in ignoring them he gives an illusory quality to his own proud position. There is no woman contributor to this volume, and one addition for the sake of appearances would hardly have done. Do women not fit into our conception of change and challenge? Is their experience of change of a character so unsensational that man's interest is not aroused? Or are their experiences of a physical and emotional nature so foreign to men that rather than contemplate them men would prefer to take off to some orbit? I mean, of course, the physical changes and the emotional challenges of that everyday miracle, the creation, the growth, and the emergence of new individuals. We habitually ascribe man's survival to the proud coherence of the schemes of men, not remembering the fact that while each scheme was tested and many exploded, women met the challenge of keeping some essentials together, of rebuilding, and of bringing up rebuilders. Here, too, I think that a new balance of Male and Female, of Paternal and Maternal, may well be presaged in the extraordinary changes in the relation of the sexes to each other, wherever science, technology, and genuine self-scrutiny advance. This, too, is another book, and yet a continuation of the symposium inaugurated here.

To turn from the postscript back to the preface: the writing of those additional books would most profitably be preceded by the careful study of this one.

Cambridge, Mass.   ERIK H. ERIKSON
September 1962

YOUTH:
change
and
challenge

ERIK H. ERIKSON

Youth: Fidelity and Diversity

THE SUBJECT of this paper is a certain strength inherent in the age of youth. I call it the sense of and the capacity for Fidelity. Such a strength, to me, is not a moral trait to be acquired by individual effort. Rather, I believe it to be part of the human equipment evolved with socio-genetic evolution. This assertion I could not undertake to defend here; nor could I make plausible the fact that, in the schedule of individual growth, Fidelity could not mature earlier in life and must not, in the crises of youth, fail its time of ascendance if human adaptation is to remain intact. Nor (to complete the list of limitations) could I review the other stages of life and the specific strengths and weaknesses contributed by each to man's precarious adaptation. We can take only a brief look at the stage of life which immediately precedes youth, the school age, and then turn to youth itself. I regret this; for even as one can understand oneself only by looking at *and* away from oneself, one can recognize the meaning of a stage only by studying it in the context of all the others.

The school age, which intervenes between childhood and youth, finds the child, previously dominated by play, ready, willing, and able to apply himself to those rudimentary skills which form the necessary preparation for his culture's tools and weapons, symbols and concepts. Also, it finds him eager to realize actual roles (previously play-acted) which promise him eventual recognition within the specializations of his culture's technology. I would say, then, that Skillfulness is the specific strength emerging in man's school age. However, the stage-by-stage acquisition during individual childhood of each of man's evolutionary gains leaves the mark of infantile experience on his proudest achievements. As the play age bequeaths to all methodical pursuits a quality of grandiose delusion, the school age leaves man with a naive acceptance of "what works."

1

As the school child makes methods his own, he also permits accepted methods to make him their own. To consider as good only what works, and to feel accepted only if things work, to manage and to be managed, can become his dominant delight and value. And since technological specialization is an intrinsic part of the human horde's or tribe's or culture's system and world image, man's pride in the tools that work with materials and animals extends to the weapons which work against other humans as well as against other species. That this can awaken a cold cunning as well as an unmeasured ferocity rare in the animal world is, of course, due to a combination of developments. Among these we will be most concerned (because it comes to the fore during youth) with man's need to combine technological pride with a sense of identity: a double sense of personal self-sameness slowly accrued from infantile experiences and of shared sameness experienced in encounters with a widening part of the community.

This need too is an evolutionary necessity as yet to be understood and influenced by planning: for men—not being a natural species any more, and not a mankind as yet—need to feel that they are of some special kind (tribe or nation, class or caste, family, occupation, or type), whose insignia they will wear with vanity and conviction, and defend (along with the economic claims they have staked out for their kind) against the foreign, the inimical, the not-so-human kinds. Thus it comes about that they can use all their proud skills and methods most systematically against other men, even in the most advanced state of rationality and civilization, with the conviction that they could not morally afford not to do so.

It is not our purpose, however, to dwell on the easy perversion and corruptibility of man's morality, but to determine what those core virtues are which—at this stage of psychosocial evolution—need our concerted attention and ethical support; for antimoralists as well as moralists easily overlook the bases in human nature for a strong ethics. As indicated, Fidelity is that virtue and quality of adolescent ego strength which belongs to man's evolutionary heritage, but which—like all the basic virtues—can arise only in the interplay of a life stage with the individuals and the social forces of a true community.

At this point, it may be necessary to defend the use of the word "virtue" in this context. It once had the connotation of an inherent strength and of an active quality in something to be described: a medicine or a drink, for example, was said to be "without virtue" when it had lost its spirit. In this sense, I think, one may use the term "basic virtues" to connote certain qualities which begin to animate man pervasively during successive stages of his life, Hope being the first and the most basic.[1] The use of such a term, however, for the conceptualization of a quality emerging from the interplay of individual growth and social structure calls to mind dreaded "naturalist" fallacies. All I can say here is that newer concepts of environment (such as the *Umwelt* of the ethologists) imply an optimum relation of inborn potentialities and the structure of the environment. This is not to deny the special problems adhering to the fact that man creates his environment and both lives in it and judges his own modes of living.

The evidence in young lives of the search for something and somebody to be true to is seen in a variety of pursuits more or less sanctioned by society. It is often hidden in a bewildering combination of shifting devotion and sudden perversity, sometimes more devotedly perverse, sometimes more perversely devoted. Yet, in all youth's seeming shiftiness, a seeking after some durability in change can be detected, whether in the accuracy of scientific and technical method or in the sincerity of conviction; in the veracity of historical and fictional accounts or the fairness of the rules of the game; in the authenticity of artistic production (and the high fidelity of reproduction) or in the genuineness of personalities and the reliability of commitments. This search is easily misunderstood, and often it is only dimly perceived by the individual himself, because youth, always set to grasp both diversity in principle and principle in diversity, must often test extremes before settling on a considered course. These extremes, particularly in times of ideological confusion and widespread marginality of identity, may include not only rebellious but also deviant, delinquent, and self-destructive tendencies. However, all this can be in the nature of a moratorium, a period of delay, in which to test the rock-bottom of some truth before committing the powers of body and mind to a segment of the existing (or a coming) order. "Loyal" and "legal" have the same root, linguistically and psychologically; for legal commitment is an unsafe burden unless shouldered with a sense of sovereign choice and experienced as loyalty. To develop that sense is a joint task of the consistency of individual life history and the ethical potency of the historical process.

Let a great tragic play tell us something of the elemental nature of the crisis man encounters here. If it is a prince's crisis, let us not

3

forget that the "leading families" of heaven and history at one time personified man's pride and tragic failure. Prince Hamlet is in his twenties, some say early, some late. We will say he is in the middle of his third decade, a youth no longer young and about to forfeit his moratorium. We find him in a tragic conflict in which he cannot make the one step demanded simultaneously by his age and his sex, his education, and his historical responsibility.

If we want to make Shakespeare's insight into one of "the ages of man" explicit, we know that such an endeavor seems reprehensible to the students of drama, if undertaken by a trained psychologist. Everybody else (how could he do otherwise?) interprets Shakespeare in the light of some prevailing if naive psychology. I will not try to solve the riddle of Hamlet's inscrutable nature, because his inscrutability is his nature. I feel sufficiently warned by Shakespeare himself, who lets Polonius speak like the caricature of a psychiatrist:

> And I do think—or else this brain of mine
> Hunts not the trail of policy so sure
> as it has us'd to do—that I have found
> The very cause of Hamlet's lunacy.

Hamlet's decision to play insane is a secret which the audience shares with him from the start, without their ever getting rid of the feeling that he is on the verge of slipping into the state he pretends. "His madness," says T. S. Eliot, "is less than madness, and more than feigned."

If Hamlet's madness is more than feigned, it appears to be aggravated at least fivefold: by habitual melancholy, an introverted personality, Danishness, an acute state of mourning, and love. All this makes a regression to the Oedipus complex, postulated by Ernest Jones as the main theme of this as of other great tragedies, entirely plausible.[2] This would mean that Hamlet cannot forgive his mother's recent illegitimate betrayal, because he had not been able as a child to forgive her for having betrayed him quite legitimately with his father; but, at the same time, he is unable to avenge his father's recent murder, because as a child he had himself betrayed him in phantasy and wished him out of the way. Thus he forever postpones—until he ruins the innocent with the guilty—his uncle's execution, which alone would free the ghost of his beloved father from the fate of being,

> doomed for a certain term to walk the night
> and for the day confined to fast in fires.

4

No audience, however, can escape the feeling that he is a man of superior conscience, advanced beyond the legal concepts of his time, consumed by his own past and by that of his society.

One further suggestion is inescapable, that Hamlet displays some of the playwright's and the actor's personality: for where others lead men and change the course of history, he reflectively moves characters about on the stage (the play within the play); in brief, where others act, he play-acts. And indeed, Hamlet may well stand, historically speaking, for an abortive leader, a still-born rebel.

We shall return to this in another context. In the meantime, all that has been stated can only support a biographic view which concentrates on Hamlet's age and status as a young intellectual of his time: for did he not recently return from studies at Wittenberg, the hotbed of humanist corruption, his time's counterpart to Sophist Athens (and today's existentialist centers of learning)?

There are five young men in the play, all Hamlet's age mates, and all sure (or even overdefined) in their identities as dutiful sons, courtiers, and future leaders. But they are all drawn into the moral swamp of infidelity, which seeps into the fiber of all those who owe allegiance to "rotten" Denmark, drawn by the multiple intrigue which Hamlet hopes to defeat with his own intrigue: the play within the play.

Hamlet's world, then, is one of diffuse realities and fidelities. Only through the play within the play and through the madness within the insanity, does Hamlet, the actor within the play-actor, reveal the identity within the pretended identities—and the superior fidelity in the fatal pretense.

His estrangement is one of identity diffusion. His estrangement from existence itself is expressed in the famous soliloquy. He is estranged from being human and from being a man: "Man delights me not; no, nor woman either"; and estranged from love and procreation: "I say we will have no more marriage." He is estranged from the ways of his country, "though I am native here and to the manner born"; and much like our "alienated" youth, he is estranged from and describes as "alienated" the overstandardized man of his day, who "only got the tune of time and outward habit of encounter."

Yet Hamlet's single-minded and tragically doomed search for Fidelity breaks through all this. Here is the essence of the historical Hamlet, that ancient model who was a hero on the folk stage for centuries before Shakespeare modernized and eternalized him:[3]

He was loth to be thought prone to lying about any matter, and

5

wished to be held a stranger to any falsehood; and accordingly he mingled craft and candor in such a wise that, though his words did not lack truth, yet there was nothing to betoken the truth and to betray how far his keenness went.

It accords with the general diffusion of truth in Hamlet that this central theme is announced in the old fool's message to his son:

> *Polonius:* This above all: to thine own self be true
> And it must follow, as the night the day,
> Thou canst not then be false to any man.

Yet it is also the central theme of Hamlet's most passionate pronouncements, which make his madness but an adjunct to his greatness. He abhors conventional sham, and advocates genuineness of feeling:

> Seems, madam! Nay, it is; I know not "seems."
> 'Tis not alone my inky cloak, good mother,
> Nor customary suits of solemn black,
> Nor windy suspiration of forc'd breath,
> No, nor the fruitful river in the eye,
> Nor the dejected havior of the visage,
> Together with all forms, moods, shapes of grief
> That can denote me truly. These indeed seem,
> For they are actions that a man might play:
> But I have that within which passes show;
> These but the trappings and the suits of woe.

He searches for what only an elite will really understand—"honest method":

> I heard thee speak me a speech once but it was never
> acted; or, if it was, not above once; for the play I
> remember, pleased not the million. . . ! it was (as I
> received it, and others, whose judgments cried in
> the top of mine) an excellent play, well digested
> and in the scenes, set down with as much modesty and
> cunning. I remember one said there were no sallets
> in the lines to make the matter savoury, nor no matter
> in the phrase that might indict the author of affection;
> but called it an honest method.

He fanatically insists on purity of form and fidelity of reproduction:

> . . . let your discretion be your tutor. Suit the
> action to the word, the word to the action, with this
> special observance, that you o'erstep not the modesty
> of nature; for anything so overdone is from the purpose
> of playing whose end, both at the first and now, was,

and is to hold, as 'twere, the mirror up to nature,
to show virtue her own image and the very age and
body of time his own form and pressure.

And finally, the eager (and overeager) acknowledgment of genuine character in his friend:

Since my dear soul was mistress of her choice
And could of men distinguish, her election
Hath sealed thee for herself; for thou hast been
As one in suffering all, that suffers nothing,
A man that fortune buffets and rewards
Hast ta'en with equal thanks; and bless'd are those
Whose blood and judgement are so co-mingled
That they are not a pipe for fortune's finger
To sound what stop she please. Give me that man
That is nor passion's slave, and I will wear him
in my heart's core, ay in my heart of heart,
As I do thee. Something too much of this.

This, then, is the Hamlet within Hamlet. It fits the combined play-actor, the intellectual, the youth, and the neurotic that his words are his better deeds, that he can say clearly what he cannot live, and that his fidelity must bring doom to those he loves: for what he accomplishes at the end is what he tried to avoid, even as he realizes what we would call his negative identity in becoming exactly what his own ethical sense could not tolerate: a mad revenger. Thus do inner reality and historical actuality conspire to deny tragic man the positive identity for which he seems exquisitely chosen. Of course, the audience all along has sensed in Hamlet's very sincerity an element of deadliness. At the end he gives his "dying voice" to his counterplayer on the historical stage, victorious young Fortinbras, who in turn insists on having him,

. . . born like a soldier to the stage
For he was likely, had he been put on,
To have prov'd most royal.

The ceremonial fanfares, blaring and hollow, announce the end of this singular youth. He is confirmed by his chosen peers, with the royal insignia of his birth. A special person, intensely human, is buried—a member of his special kind.

To be a special kind, we have said, is an important element in the human need for personal and collective identities—all, in a

7

sense, pseudospecies. They have found a transitory fulfillment in man's greatest moments of cultural identity and civilized perfection, and each such tradition of identity and perfection has highlighted what man could be, could he be all these at one time. The utopia of our own era predicts that man will be one species in one world, with a universal identity to replace the illusory superidentities which have divided him, and with an international ethics replacing all moral systems of superstition, repression, and suppression. Whatever the political arrangement that will further this utopia, we can only point to the schedule of human strengths which potentially emerge with the stages of life and indicate their interdependence on the structure of communal life. In youth, ego strength emerges from the mutual confirmation of individual and community, in the sense that society recognizes the young individual as a bearer of fresh energy and that the individual so confirmed recognizes society as a living process which inspires loyalty as it receives it, maintains allegiance as it attracts it, honors confidence as it demands it.

Let us go back, then, to the origins of that combination of drivenness and disciplined energy, of irrationality and courageous capability which belong to the best discussed and the most puzzling phenomena of the life cycle. The puzzle, we must grant throughout, is in the essence of the phenomenon. For the unity of the personality must be unique to be united, and the functioning of each new generation unpredictable to fulfill its function.

Of the three sources of new energy, physical growth is the most easily measured and systematically exercised, although its contribution to the aggressive drives is little understood. The youthful powers of comprehension and cognition can be experimentally studied and with planning applied to apprenticeship and study, but their relation to ideological imagination is less well known. Finally, the long delayed genital maturation is a source of untold energy, but also of a drivenness accompanied by intrinsic frustration.

When maturing in his physical capacity for procreation, the human youth is as yet unable to love in that binding manner which only two identities can offer each other; nor to care consistently enough to sustain parenthood. The two sexes, of course, differ greatly in these respects, and so do individuals, while societies provide different opportunities and sanctions within which individuals must fend for their potentials—and for their potency. But what I have called a psychosocial moratorium, of some form and duration between the advent of genital maturity and the onset of responsible

8

adulthood, seems to be built into the schedule of human development. Like all the moratoria in man's developmental schedules, the delay of adulthood can be prolonged and intensified to a forceful and a fateful degree; thus it accounts for very special human achievements and also for the very special weaknesses in such achievements. For, whatever the partial satisfactions and partial abstinences that characterize premarital sex life in various cultures—whether the pleasure and pride of forceful genital activity without commitment, or of erotic states without genital consummation, or of disciplined and devoted delay—ego development uses the psychosexual powers of adolescence for enhancing a sense of style and identity. Here, too, man is never an animal: even where a society furthers the genital closeness of the sexes, it does so in a stylized manner. On the other hand, the sex act, biologically speaking, is the procreative act, and there is an element of psychobiological dissatisfaction in any sexual situation not favorable in the long run to procreative consummation and care—a dissatisfaction which can be tolerated by otherwise healthy people, as all partial abstinences can be borne: for a certain period, under conditions otherwise favorable to the aims of identity formation. In the woman, no doubt, this dissatisfaction plays a much greater role, owing to her deeper engagement, physiologically and emotionally, in the sex act as the first step in a procreative commitment of which her monthly cycle is a regular bodily and emotive reminder.

The various hindrances to a full consummation of adolescent genital maturation have many deep consequences for man which pose an important problem for future planning. Best known is the regressive revival of that earlier stage of psychosexuality which preceded even the emotionally quiet first school years, that is, the infantile genital and locomotor stage, with its tendency toward auto-erotic manipulation, grandiose phantasy, and vigorous play.[4] But in youth, auto-erotism, grandiosity, and playfulness are all immensely amplified by genital potency and locomotor maturation, and are vastly complicated by what we will presently describe as the youthful mind's new historical perspective.

The most widespread expression of the discontented search of youth is the craving for locomotion, whether expressed in a general "being on the go," "tearing after something," or "running around"; or in locomotion proper, as in vigorous work, in absorbing sports, in rapt dancing, in shiftless *Wanderschaft*, and in the employment and misuse of speedy animals and machines. But it also finds expression through participation in the movements of the day (whether the riots of a local commotion or the parades and campaigns of major ideological forces), if they only appeal to the need for feeling "moved" and for feeling essential in moving something along toward an open future. It is clear that societies offer any number of ritual combinations of ideological perspective and vigorous movement (dance, sports, parades, demonstrations, riots) to harness youth in the service of their historical aims; and that where societies fail to do so, these patterns will seek their own combinations, in small groups occupied with serious games, good-natured foolishness, cruel prankishness, and delinquent warfare. In no other stage of the life cycle, then, are the promise of finding oneself and the threat of losing oneself so closely allied.

In connection with locomotion, we must mention two great industrial developments; the motor engine and the motion picture. The motor engine, of course, is the very heart and symbol of our technology and its mastery, the aim and aspiration of much of modern youth. In connection with immature youth, however, it must be understood that both motor car and motion pictures offer to those so inclined passive locomotion with an intoxicating delusion of being intensely active. The prevalence of car thefts and motor accidents among juveniles is much decried (although it is taking the public a long time to understand that a theft is an appropriation for the sake of gainful possession), while automobiles more often than not are stolen by the young in search of a kind of automotive intoxication, which may literally run away with car and youngster. Yet, while vastly inflating a sense of motor omnipotence, the need for active locomotion often remains unfulfilled. Motion pictures especially offer the onlooker, who sits, as it were, with the engine of his emotions racing, fast and furious motion in an artificially widened visual field, interspersed with close-ups of violence and sexual possession—and all this without making the slightest demand on intelligence, imagination, or effort. I am pointing here to a widespread imbalance in adolescent experience, because I think it explains new kinds of adolescent outbursts and points to new necessities of mastery. The danger involved is greatly balanced in that part of youth which can take active charge of technical development, manages to learn, and to identify with the ingeniousness of invention, the improvement of production and the care of machinery, and is thus offered a new and unlimited application of youthful capacities. Where youth is underprivileged in such technical experience, it must explode in riotous motion; where it is ungifted, it will feel estranged from the modern world, until technology and nontechnical intelligence have come to a certain convergence.

The cognitive gifts developing during the first half of the second decade add a powerful tool to the tasks of youth. J. Piaget calls the gains in cognition made toward the middle teens, the achievement of "formal operations."[5] This means that the youth can now operate on hypothetical propositions, can think of possible variables and potential relations, and think of them in thought alone, independent of certain concrete checks previously necessary. As Jerome S. Bruner puts it, the child now can "conjure up systematically the full range of alternative possibilities that could exist at any given time."[6] Such cognitive orientation forms not a contrast but a complement to the need of the young person to develop a sense of identity, for, from among all possible and imaginable relations, he must make a series of ever narrowing selections of personal, occupational, sexual, and ideological commitments.

Here again diversity and fidelity are polarized: they make each other significant and keep each other alive. Fidelity without a sense of diversity can become an obsession and a bore; diversity without a sense of fidelity, an empty relativism.

The sense of ego identity, then, becomes more necessary (and more problematical) wherever a wide range of possible identities is envisaged. Identity is a term used in our day with faddish ease; at this point, I can only indicate how very complicated the real article is.[7] For ego identity is partially conscious and largely unconscious. It is a psychological process reflecting social processes; but with sociological means it can be seen as a social process reflecting psychological processes; it meets its crisis in adolescence, but has grown throughout childhood and continues to re-emerge in the crises of later years. The overriding meaning of it all, then, is the creation of a sense of sameness, a unity of personality now felt by the individual and recognized by others as having consistency in time— of being, as it were, an irreversible historical fact.

The prime danger of this age, therefore, is identity confusion, which can express itself in excessively prolonged moratoria (Hamlet offers an exalted example); in repeated impulsive attempts to end the moratorium with sudden choices, that is, to play with historical possibilities, and then to deny that some irreversible commitment has already taken place; and sometimes also in severe regressive pathology, which we will illustrate presently. The dominant issue of

11

this, as of any other stage, therefore, is that of the active, the selective, ego being in charge and being enabled to be in charge by a social structure which grants a given age group the place it needs —and in which it is needed.

In a letter to Oliver Wendell Holmes, William James speaks of wanting to "rebaptize himself" in their friendship—and this one word says much of what is involved in the radical direction of the social awareness and the social needs of youth. From the middle of the second decade, the capacity to think and the power to imagine reach beyond the persons and personalities in which youth can immerse itself so deeply. Youth loves and hates in people what they "stand for" and chooses them for a significant encounter involving issues that often, indeed, are bigger than you and I. We have heard Hamlet's declaration of love to his friend Horatio, a declaration quickly broken off—"something too much here." It is a new reality, then, for which the individual wishes to be reborn, with and by those whom he chooses as his new ancestors and his genuine contemporaries.

This mutual selection, while frequently associated with, and therefore interpreted as a rebellion against or withdrawal from, the childhood environment, is an expression of a truly new perspective which I have already called "historical"—in one of those loose uses of an ancient and overspecialized word which sometimes become necessary in making new meanings specific. I mean by "historical perspective" something which every human being newly develops during adolescence. It is a sense of the irreversibility of significant events and an often urgent need to understand fully and quickly what kind of happenings in reality and in thought determine others, and why. As we have seen, psychologists such as Piaget recognize in youth the capacity to appreciate that any process can be understood when it is retraced in its steps and thus reversed in thought. Yet it is no contradiction to say that he who comes to understand such a reversal also realizes that in reality, among all the events that can be thought of, a few will determine and narrow one another with historical fatality, whether (in the human instance) deservedly or undeservedly, intentionally or unintentionally.

Youth, therefore, is sensitive to any suggestion that it may be hopelessly determined by what went before in life histories or in history. Psychosocially speaking, this would mean that irreversible childhood identifications would deprive an individual of an identity of his own; historically, that invested powers should prevent a group from realizing its composite historical identity. For these reasons, youth often rejects parents and authorities and wishes to belittle them as inconsequential; it is in search of individuals and movements who claim, or seem to claim, that they can predict what is irreversible, thus getting ahead of the future—which means, reversing it. This in turn accounts for the acceptance by youth of mythologies and ideologies predicting the course of the universe or the historical trend; for even intelligent and practical youth can be glad to have the larger framework settled, so that it can devote itself to the details which it can manage, once it knows (or is convincingly told) what they stand for and where it stands. Thus, "true" ideologies are verified by history—for a time; for, if they can inspire youth, youth will make the predicted history come more than true.

By pointing to what, in the mind of youth, people "stand for," I did not mean to overemphasize the ideological explicitness in the meaning of individuals to youth. The selection of meaningful individuals can take place in the framework of pointed practicalities such as schooling or job selection, as well as in religious and ideological fellowship; while the methods of selection can range from banal amenity and enmity to dangerous play with the borderlines of sanity and legality. But the occasions have in common a mutual sizing up and a mutual plea for being recognized as individuals who can be more than they seem to be, and whose potentials are needed by the order that is or will be. The representatives of the adult world thus involved may be advocates and practitioners of technical accuracy, of a method of scientific inquiry, of a convincing rendition of truth, of a code of fairness, of a standard of artistic veracity, or of a way of personal genuineness. They become representatives of an elite in the eyes of the young, quite independently of whether or not they are also viewed thus in the eyes of the family, the public, or the police. The choice can be dangerous, but to some youths the danger is a necessary ingredient of the experiment. Elemental things are dangerous; and if youth could not overcommit itself to danger, it could not commit itself to the survival of genuine values—one of the primary steering mechanisms of psychosocial evolution. The elemental fact is that only when fidelity has found its field of manifestation is the human as good as, say, the nestling in nature, which is ready to rely on its own wings and to take its adult place in the ecological order.

If in human adolescence this field of manifestation is alternately one of devoted conformism and of extreme deviancy, of rededication and of rebellion, we must remember the necessity for man to react (and to react most intensively in his youth) to the diversity of conditions. In the setting of psychosocial evolution, we can ascribe a long-range meaning to the idiosyncratic individualist and to the rebel as well as to the conformist, albeit under different historical conditions. For healthy individualism and devoted deviancy contain an indignation in the service of a wholeness that is to be restored, without which psychosocial evolution would be doomed. Thus, human adaptation has its loyal deviants, its rebels, who refuse to adjust to what so often is called, with an apologetic and fatalistic misuse of a once good phrase, "the human condition."

Loyal deviancy and identity formation in extraordinary individuals are often associated with neurotic and psychotic symptoms, or at least with a prolonged moratorium of relative isolation, in which all the estrangements of adolescence are suffered. In *Young Man Luther* I have attempted to put the suffering of a great young man into the context of his greatness and his historic position.[8]

It is not our purpose, however, to discuss what to many youths is the most urgent question, and yet to us the most difficult to answer, namely, the relation of special giftedness and neurosis; rather, we must characterize the specific nature of adolescent psychopathology, or, even more narrowly, indicate the relevance of the issue of fidelity to the psychopathology of youth.

In the classical case of this age group, Freud's first published encounter with an eighteen-year-old girl suffering from "*petite hystérie* with the commonest of all . . . symptoms," it is interesting to recall that at the end of treatment Freud was puzzled as to "what kind of help" the girl wanted from him. He had communicated to her his interpretation of the structure of her neurotic disorder, an interpretation which became the central theme of his classical publication on the psychosexual factors in the development of hysteria.[9] Freud's clinical reports, however, remain astonishingly fresh over the decades, and today his case history clearly reveals the psychosocial centering of the girl's story in matters of fidelity. In fact, one might say, without seriously overdoing it, that three words characterize her social history: sexual infidelity on the part of some of the most important adults in her life; the perfidy of her father's denial of his friend's sexual acts, which were in fact the precipitating cause of the girl's illness; and a strange tendency on the part of all the adults around the girl to make her a confidante in any number of matters, without having enough confidence in her to acknowledge the truths relevant to her illness.

Freud, of course, focused on other matters, opening up, with the concentration of a psychosurgeon, the symbolic meaning of her symptoms and their history; but, as always, he reported relevant data on the periphery of his interests. Thus, among the matters which somewhat puzzled him, he reports that the patient was "almost beside herself at the idea of its being supposed that she had merely fancied" the conditions which had made her sick; and that she was kept "anxiously trying to make sure whether I was being quite straightforward with her"—or perfidious like her father. When at the end she left analyst and analysis "in order to confront the adults around her with the secrets she knew," Freud considered this an act of revenge on them, and on him; and within the outlines of his interpretation, this partial interpretation stands. Nevertheless, as we can now see, there was more to this insistence on the historical truth than the denial of an inner truth—and this especially in an adolescent. For, the question as to what confirms them irreversibly as a truthful or a cheating, a sick or a rebellious type is paramount in the minds of adolescents; and the further question, whether or not they were right in not accepting the conditions which made them sick, is as important to them as the insight into the structure of their sickness can ever be. In other words, they insist that the meaning of their sickness find recognition within a reformulation of the historical truth as revealed in their own insights and distortions, and not according to the terms of the environment which wishes them to be "brought to reason" (as Dora's father had put it, when he brought her to Freud).

No doubt, Dora by then was a hysteric, and the meaning of her symptoms was psychosexual; but the sexual nature of her disturbance and of the precipitating events should not blind us to the fact that other perfidies, familial and communal, cause adolescents to regress in a variety of ways to a variety of earlier stages.

Only when adolescence is reached does the capacity for such clear regression and symptom formation occur: only when the historical function of the mind is consolidated can significant repressions become marked enough to cause consistent symptom formation and deformation of character. The depth of regression determines the nature of the pathology and points to the therapy to be employed. However, there is a pathognomic picture which all sick youth have in common and which is clearly discernible in Freud's description of Dora's total state. This picture is characterized first of all by a denial of the historical flux of time, and by an attempt to challenge retrospectively, while retesting in the present all

15

parental premises before new trust is invested in the (emancipated) future.

The sick adolescent thus gradually stops extending experimental feelers toward the future; his moratorium of illness becomes an end in itself and thus ceases to be a moratorium (Dora suffered from a *"taedium vitae* which was probably not entirely genuine," Freud wrote). It is for this reason that death and suicide can be at this time such a spurious preoccupation—one leading unpredictably to suicide (and to murder)—for death would conclude the life history before it joins others in inexorable commitment (Dora's parents found "a letter in which she took leave of them because she could no longer endure life. Her father . . . guessed that the girl had no serious suicidal intentions.") There is also a social isolation which excludes all sense of solidarity and can lead to a snobbish isolation which finds companions but no friends (Dora "tried to avoid social intercourse," was "distant" and "unfriendly"). The energy of repudiation which accompanies the first steps of an identity formation (and in some youngsters can lead to the sudden impulse to annihilate) is in neurotics turned against the self ("Dora was satisfied neither with herself nor with her family").

A repudiated self in turn cannot offer loyalty, and, of course, fears the fusion of love or of sexual encounters. The work inhibition often connected with this picture (Dora suffered from "fatigue and lack of concentration") is really a career inhibition, in the sense that every exertion of skill or method is suspected of binding the individual to the role and the status suggested by the activity; thus, again, any moratorium is spoiled. Where fragmentary identities are formed, they are highly self-conscious and are immediately put to a test (thus Dora obviously defeated her wish to be a woman intellectual). This identity consciousness is a strange mixture of superority, almost a megalomania ("I am a majority of one," one of my patients said), with which the patient tries to convince himself that he is really too good for his community or his period of history, while he is equally convinced of being nobody.

We have sketched the most obvious social symptoms of adolescent psychopathology, in part to indicate that, besides the complicated structure of specific symptoms, there is in the picture presented of each stage an expression of the dominant psychosocial issue, so open that one sometimes wonders whether the patient lies by telling the simple truth or tells the truth when he seems most obviously to lie.

The sketch presented, however, also serves as a comparison of the isolated adolescent sufferer with those youths who try to solve their doubt in their elders by joining deviant cliques and gangs. Freud found that "psychoneuroses are, so to speak, the negative of perversions,"[10] which means that neurotics suffer under the repression of tendencies which perverts try to "live out." This has a counterpart in the fact that isolated sufferers try to solve by withdrawal what the joiners of deviant cliques and gangs attempt to solve by conspiracy.

If we now turn to this form of adolescent pathology, the denial of the irreversibility of historical time appears to be expressed in a clique's or a gang's delusion of being an organization with a tradition and an ethics all its own. The pseudo-historical character of such societies is expressed in such names as "The Navahos," "The Saints," or "The Edwardians"; while their provocation is countered by society (remember the Pachucos of the war years) with a mixture of impotent rage wherever murderous excess does actually occur, and with a phobic overconcern followed by vicious suppression wherever these "secret societies" are really no more than fads lacking any organized purpose. Their pseudo-societal character reveals itself in their social parasitism, and their pseudo-rebellion in the conformism actually governing their habits. Yet the seemingly unassailable inner sense of callous rightness is no doubt due to an inner realignment of motivations, which can best be understood by briefly comparing the torment of the isolated youngster with the temporary gains derived by the joiner from the mere fact that he has been taken into a pseudo-society. The time diffusion attending the isolate's inability to envisage a career is "cured" by his attention to "jobs"—theft, destruction, fights, murder, or acts of perversion or addiction, conceived on the spur of the moment and executed forthwith. This "job" orientation also takes care of the work inhibition, because the clique and the gang are always "busy," even if they just "hang around." Their lack of any readiness to wince under shaming or accusation is often considered the mark of a total personal perdition, while in fact it is a trademark, an insignia of the "species" to which the youngster (mostly marginal in economic and ethnic respects) would rather belong than to a society which is eager to confirm him as a criminal and then promises to rehabilitate him as an ex-criminal.

As to the isolate's tortured feelings of bisexuality or of an immature need for love, the young joiner in social pathology, by joining, has made a clear decision: he is male with a vengeance, she, a female without sentimentality; or they are both perverts. In either case, they can eliminate the procreative function of genitality altogether and can make a pseudo-culture of what is left. By the same token, they will acknowledge authority only in the form chosen in the act of joining, repudiating the rest of the social world, where the isolate repudiates existence as such and, with it, himself.

The importance of these comparative considerations, which have been stated in greater detail elsewhere, lie in the impotent craving of the isolated sufferer to be true to himself, and in that of the joiner, to be true to a group and to its insignia and codes. By this I do not mean to deny that the one is sick (as his physical and mental symptoms attest), nor that the other can be on the way to becoming a criminal, as his more and more irreversible acts and choices attest. Both theory and therapy, however, lack the proper leverage, if the need for (receiving and giving) fidelity is not understood, and especially if instead the young deviant is confirmed by every act of the correctional or therapeutic authorities as a future criminal or a lifelong patient.

In Dora's case, I have tried to indicate the phenomenology of this need. As to young delinquents, I can only quote again one of those rare newspaper reports which convey enough of a story to show the elements involved. Kai T. Erikson and I have used this example as an introduction to our article "The Confirmation of the Delinquent."[11]

JUDGE IMPOSES ROAD GANG TERM FOR BACK TALK

Wilmington, N. D. (UP)—A "smart alecky" youth who wore pegged trousers and a flattop haircut began six months on a road gang today for talking back to the wrong judge.

Michael A. Jones, 20, of Wilmington, was fined $25 and costs in Judge Edwin Jay Roberts Jr.'s superior court for reckless operation of an automobile. But he just didn't leave well enough alone.

"I understand how it was, with your pegged trousers and flattop haircut," Roberts said in assessing the fine. "You go on like this and I predict in five years you'll be in prison."

When Jones walked over to pay his fine, he overheard Probation Officer Gideon Smith tell the judge how much trouble the "smart alecky" young offender had been.

"I just want you to know I'm not a thief," interrupted Jones to the judge.

The judge's voice boomed to the court clerk: "Change that judgment to six months on the roads."

I quote the story here to add the interpretation that the judge in

Youth: Fidelity and Diversity

this case (neither judge nor case differs from a host of others) took it as an affront to the dignity of authority what may have also been a desperate "historical" denial, an attempt to claim that a truly antisocial identity had not yet been formed, and that there was enough discrimination and potential fidelity left to be made something of by somebody who cared to do so. But instead, what the young man and the judge made of it was likely, of course, to seal the irreversibility and confirm the doom. I say "was likely to," because I do not know what happened in this case; we do know, however, the high recidivity of criminality in the young who, during the years of identity formation, are forced by society into intimate contact with criminals.

Finally, it cannot be overlooked that at times political undergrounds of all kinds can and do make use of the need for fidelity as well as the store of wrath in those deprived in their need by their families or their societies. Here social rejuvenation can make use of and redeem social pathology, even as in individuals special giftedness can be related to and redeem neurosis. These are matters too weighty to be discussed briefly and, at any rate, our concern has been with the fact that the psychopathology of youth suggests a consideration of the same issues which we found operative in the evolutionary and developmental aspects of this stage of life.

To summarize: Fidelity, when fully matured, is the strength of disciplined devotion. It is gained in the involvement of youth in such experiences as reveal the essence of the era they are to join— as the beneficiaries of its tradition, as the practitioners and innovators of its technology, as renewers of its ethical strength, as rebels bent on the destruction of the outlived, and as deviants with deviant commitments. This, at least, is the potential of youth in psychosocial evolution; and while this may sound like a rationalization endorsing any high-sounding self-delusion in youth, any self-indulgence masquerading as devotion, or any righteous excuse for blind destruction, it makes intelligible the tremendous waste attending this as any other mechanism of human adaptation, especially if its excesses meet with more moral condemnation than ethical guidance. On the other hand, our understanding of these processes is not furthered by the "clinical" reduction of adolescent phenomena to their infantile antecedents and to an underlying dichotomy of drive and conscience. Adolescent development comprises a new set of identification processes, both with significant persons and with ideological forces,

19

which give importance to individual life by relating it to a living community and to ongoing history, and by counterpointing the newly won individual identity with some communal solidarity.

In youth, then, the life history intersects with history: here individuals are confirmed in their identities, societies regenerated in their life style. This process also implies a fateful survival of adolescent modes of thinking in man's historical and ideological perspectives.

Historical processes, of course, have already entered the individual's core in childhood. Both ideal and evil images and the moral prototypes guiding parental administrations originate in the past struggles of contending cultural and national "species," which also color fairytale and family lore, superstition and gossip, and the simple lessons of early verbal training. Historians on the whole make little of this; they describe the visible emergence and the contest of autonomous historical ideas, unconcerned with the fact that these ideas reach down into the lives of generations and re-emerge through the daily awakening and training of historical consciousness in young individuals.

It is youth, then, which begins to develop that sense of historical irreversibility which can lead to what we may call acute historical estrangement. This lies behind the fervent quest for a sure meaning in individual life history and in collective history, and behind the questioning of the laws of relevancy which bind datum and principle, event and movement. But it is also, alas, behind the bland carelessness of that youth which denies its own vital need to develop and cultivate a historical consciousness—and conscience.

To enter history, each generation of youth must find an identity consonant with its own childhood and consonant with an ideological promise in the perceptible historical process. But in youth the tables of childhood dependence begin slowly to turn: no longer is it merely for the old to teach the young the meaning of life, whether individual or collective. It is the young who, by their responses and actions, tell the old whether life as represented by the old and as presented to the young has meaning; and it is the young who carry in them the power to confirm those who confirm them and, joining the issues, to renew and to regenerate, or to reform and to rebel.

I will not at this point review the institutions which participate in creating the retrospective and the prospective mythology offering historical orientation to youth: obviously, the mythmakers of religion and politics, the arts and the sciences, the stage and fiction—all contribute to the historical logic preached to youth more or less consciously, more or less responsibly. And today we must add, at least in the United States, psychiatry; and all over the world, the press, which forces the leaders to make history in the open and to accept reportorial distortion as a major historical factor.

I have spoken of Hamlet as an abortive ideological leader. His drama combines all the elements of which successful ideological leaders are made: they are the postadolescents who make out of the very contradictions of adolescence the polarities of their charisma. Individuals with an uncommon depth of conflict, they also have uncanny gifts, and often uncanny luck with which they offer to the crisis of a generation the solution of their own crisis—always, as Woodrow Wilson put it, being "in love with activity on a large scale," always feeling that their one life must be made to count in the lives of all, always convinced that what they felt as adolescents was a curse, a fall, an earthquake, a thunderbolt, in short, a revelation to be shared with their generation and with many to come. Their humble claim to being chosen does not preclude a wish to universal power. "Fifty years from now," wrote Kierkegaard in the journal of his spiritual soliloquy, "the whole world will read my diary." He sensed, no doubt, that the impending dominance of mass ideologies would bring to the fore his cure for the individual soul, existentialism. We must study the question (I have approached it in my study of young Luther) of what ideological leaders do to history—whether they first aspire to power and then face spiritual qualms, or first face spiritual perdition and then seek universal influence. Their answers often manage to subsume under the heading of a more embracing identity all that ails man, especially young man, at critical times: danger from new weapons and from natural forces aggravated by man's misuse of nature; anxiety from sources within the life-history typical for the time; and existential dread of the ego's limitations, magnified in times of disintegrating superidentities and intensified in adolescence.

But does it not take a special and, come to think of it, a strange sense of calling, to dare and to care to give such inclusive answers? Is it not probable and in fact demonstrable that among the most passionate ideologists there are unreconstructed adolescents, transmitting to their ideas the proud moment of their transient ego recovery, of their temporary victory over the forces of existence and history, but also the pathology of their deepest isolation, the defensiveness of their forever adolescing egos—and their fear of the calm of adulthood? "To live beyond forty," says Dostoevsky's under-

21

ground diarist, "is bad taste." It warrants study, both historical and psychological, to see how some of the most influential leaders have turned away from parenthood, only to despair in middle age of the issue of their leadership as well.

It is clear that today the ideological needs of all but intellectual youth of the humanist tradition are beginning to be taken care of by a subordination of ideology to technology: what works, on the grandest scale, is good. It is to be hoped that the worst implications of this trend have outlived themselves already in fascism. Yet, in the technological superidentity, the American dream and the Marxist revolution also meet. If their competition can be halted before mutual annihilation, it is just possible that a new mankind, seeing that it can now build and destroy anything it wishes, will focus its intelligence (feminine as well as masculine) on the ethical question concerning the workings of human generations—beyond products, powers, and ideas. Ideologies in the past have contained an ethical corrective, but ethics must eventually transcend ideology as well as technology: the great question will be and already is, what man, on ethical grounds and without moralistic self-destruction, must decide *not* to do, even though he could make it work—for a while.

Moralities sooner or later outlive themselves, ethics never: this is what the need for identity and for fidelity, reborn with each generation, seems to point to. Morality in the moralistic sense can be shown by modern means of inquiry to be predicated on superstitions and irrational inner mechanisms which ever again undermine the ethical fiber of generations; but morality is expendable only where ethics prevail. This is the wisdom that the words of many languages have tried to tell man. He has tenaciously clung to the words, even though he has understood them only vaguely, and in his actions has disregarded or perverted them completely. But there is much in ancient wisdom which can now become knowledge.

As in the near future peoples of different tribal and national pasts join what must become the identity of one mankind, they can find an initial common language only in the workings of science and technology. This in turn may well help them to make transparent the superstitions of their traditional moralities and may even permit them to advance rapidly through a historical period during which they must put a vain superidentity of neonationalism in the place of their much exploited historical identity weakness. But they must also look beyond the major ideologies of the now "established" world, offered them as ceremonial masks to frighten and to attract them. The overriding issue is the creation not of a new ideology but of a universal ethics growing out of a universal technological civilization. This can be advanced only by men and women who are neither ideological youths nor moralistic old men, but who know that from generation to generation the test of what you produce is in the *care* it inspires. If there is any chance at all, it is in a world more challenging, more workable, and more venerable than all myths, retrospective or prospective: it is in historical reality, at last ethically cared for.

REFERENCES

1. For an evolutionary and genetic rationale of this concept of the life cycle, see the writer's "The Roots of Virtue," in *The Humanist Frame*, Sir Julian Huxley, ed. London: Allen and Unwin, 1961; New York: Harper and Brothers, 1961. For a more detailed exposition, see the writer's forthcoming book, *Life Cycle and Community*, in which the other stages of development are treated in chapters analogous to the present discussion of youth.

2. Ernest Jones, *Hamlet and Oedipus*. New York: Doubleday, Anchor, 1949.

3. Saxo Grammaticus, *Danish History*, translated by Elton, 1894 (quoted in Jones, *op. cit.*, pp. 163-164.

4. The classical psychoanalytic works concerned with psychosexuality and the ego defenses of youth are: Sigmund Freud, *Three Essays on the Theory of Sexuality*, standard edition, (London, The Hogarth Press, 1953), vol. 7; and Anna Freud, *The Ego and the Mechanisms of Defence*, New York, International Universities Press, 1946. For the writer's views, see his *Childhood and Society*. New York: W. W. Norton, 1950.

5. B. Inhelder and J. Piaget, *The Growth of Logical Thinking from Childhood to Adolescence*. New York: Basic Books, 1958.

6. Jerome S. Bruner, *The Process of Education*. Cambridge: Harvard University Press, 1960.

7. See the writer's "The Problem of Ego-Identity" in *Identity and the Life Cycle: Psychological Issues* (New York: International Universities Press, 1959), vol. I, no. 1.

8. *Young Man Luther*. New York: W. W. Norton, 1958; London: Faber and Faber, 1959.

9. Sigmund Freud, *Fragment of an Analysis of a Case of Hysteria*, standard edition (London: The Hogarth Press, 1953), vol. 7.

10. *Ibid.*, p. 50.

11. Erik H. Erikson and Kai T. Erikson, "The Confirmation of the Delinquent," *The Chicago Review*, Winter 1957, *10*: 15-23.

S. N. EISENSTADT

Archetypal Patterns of Youth

YOUTH CONSTITUTES a universal phenomenon. It is first of all a
biological phenomenon, but one always defined in cultural terms.
In this sense it constitutes a part of a wider cultural phenomenon,
the varying definitions of age and of the differences between one
age and another.[1] Age and age differences are among the basic
aspects of life and the determinants of human destiny. Every human
being passes through various ages, and at each one he attains and
uses different biological and intellectual capacities. At each stage he
performs different tasks and roles in relation to the other members
of his society: from a child, he becomes a father; from a pupil, a
teacher; from a vigorous youth, a mature adult, and then an aging
and "old" man.

This gradual unfolding of power and capacity is not merely a
universal, biologically conditioned, and inescapable fact. Although
the basic biological processes of maturation (within the limits set
by such factors as relative longevity) are probably more or less
similar in all human societies, their cultural definition varies from
society to society, at least in details. In all societies, age serves as a
basis for defining the cultural and social characteristic of human
beings, for the formation of some of their mutual relations and
common activities, and for the differential allocation of social roles.

The cultural definitions of age and age differences contain several
different yet complementary elements. First, these definitions often
refer to the social division of labor in a society, to the criteria according to which people occupy various social positions and roles within
any society. For instance, in many societies certain roles—especially
those of married men, full citizens, independent earners—are barred
to young people, while others—as certain military roles—are specifi-

24

cally allocated to them. Second, the cultural definition of age is one important constituent of a person's self-identity, his self-perception in terms of his own psychological needs and aspirations, his place in society, and the ultimate meaning of his life.

Within any such definition, the qualities of each age are evaluated according to their relation to some basic, primordial qualities, such as vigor, physical and sexual prowess, the ability to cope with material, social, and supernatural environment, wisdom, experience, or divine inspiration. Different ages are seen in different societies as the embodiments of such qualities. These various qualities seem to unfold from one age to another, each age emphasizing some out of the whole panorama of such possible qualities. The cultural definition of an age span is always a broad definition of human potentialities, limitations, and obligations at a given stage of life. In terms of these definitions, people map out the broad contours of life, their own expectations and possibilities, and place themselves and their fellow men in social and cultural positions, ascribing to each a given place within these contours.

The various qualities attributed to different ages do not constitute an unconnected series. They are usually interconnected in many ways. The subtle dialectics between the unfolding of some qualities and the waning of others in a person is not a mere registration of his psychological or biological traits; rather, it constitutes the broad framework of his potentialities and their limits throughout his life span. The characteristics of any one "age," therefore, cannot be fully understood except in relation to those of other ages. Whether seen as a gradually unfolding continuum or as a series of sharp contrasts and opposed characteristics, they are fully explicable and understandable only in terms of one another. The boy bears within himself the seeds of the adult man; else, he must as an adult acquire new patterns of behavior, sharply and intentionally opposed to those of his boyhood. The adult either develops naturally into an old man— or decays into one. Only when taken together do these different "ages" constitute the entire map of human possibilities and limitations; and, as every individual usually must pass through them all, their complementariness and continuity (even if defined in discontinuous and contrasting terms) become strongly emphasized and articulated.

The same holds true for the age definitions of the two sexes, although perhaps with a somewhat different meaning. Each age span is defined differently for either sex, and these definitions are usually related and complementary, as the "sexual image" and identity always constitute basic elements of man's image in every society. This close connection between different ages necessarily stresses the problem of transition from one point in a person's life to another as a basic constituent of any cultural definition of an "age." Hence, each definition of age must necessarily cope with the perception of time, and changes in time, of one's own progress in time, one's transition from one period of life to another.

This personal transition, or temporal progress, or change, may become closely linked with what may be called cosmic and societal time.[2] The attempt to find some meaning in personal temporal transition may often lead to identification with the rhythms of nature or history, with the cycles of the seasons, with the unfolding of some cosmic plan (whether cyclical, seasonal, or apocalyptic), or with the destiny and development of society. The nature of this linkage often constitutes the focus round which an individual's personal identity becomes defined in cultural terms and through which personal experience, with its anguish, may be given some meaning in terms of cultural symbols and values.

The whole problem of age definition and the linkage of personal time and transition with cosmic time become especially accentuated in that age span usually designated as youth. However great the differences among various societies, there is one focal point within the life span of an individual which in most known societies is to some extent emphasized: the period of youth, of transition from childhood to full adult status, or full membership in the society. In this period the individual is no longer a child (especially from the physical and sexual point of view) but is ready to undertake many attributes of an adult and to fulfill adult roles. But he is not yet fully acknowledged as an adult, a full member of the society. Rather, he is being "prepared," or is preparing himself for such adulthood.

This image of youth—the cultural definition of youth—contains all the crucial elements of any definition of age, usually in an especially articulated way. This is the stage at which the individual's personality acquires the basic psychological mechanism of self-regulation and self-control, when his self-identity becomes crystallized. It is also the stage at which the young are confronted with some models of the major roles they are supposed to emulate in adult life and with the major symbols and values of their culture and community. Moreover, in this phase the problem of the linkage of the personal temporal transition with cosmic or societal time becomes extremely acute. Any cultural definition of youth describes it as a transitory phase, couched in terms of transition toward something new, something basically different from the past. Hence the acuteness of the problem of linkage.

The very emphasis on the transitory nature of this stage and of its essentially preparatory character, however, may easily create a somewhat paradoxical situation. It may evolve an image of youth as the purest manifestation and repository of ultimate cultural and societal values. Such an image is rooted first in the fact that to some extent youth is always defined as a period of "role moratorium," that is, as a period in which one may play with various roles without definitely choosing any. It does not yet require the various compromises inherent in daily participation in adult life. At the same time, however, since it is also the period when the maximum identification with the values of the society is stressed, under certain conditions it may be viewed as the repository of all the major human virtues and primordial qualities. It may then be regarded as the only age in which full identification with the ultimate values and symbols of the society is attained—facilitated by the flowering of physical vigor, a vigor which may easily become identified with a more general flowering of the cosmos or the society.

The fullest, the most articulate and definitive expression of these archetypal elements of youth is best exemplified in the ritual dramatization of the transition from adolescence to adulthood, such as the various *rites de passage* and ceremonies of initiation in primitive tribes and in ancient civilizations.[3] In these rites the pre-adult youth are transformed into full members of the tribe. This transformation is effected through:

1. a series of rites in which the adolescents are symbolically divested of the characteristics of youth and invested with those of adulthood, from a sexual and social point of view; this investment, which has deep emotional significance, may have various concrete manifestations: bodily mutilation, circumcision, the taking on of a new name or symbolic rebirth;

2. the complete symbolic separation of the male adolescents from the world of their youth, especially from their close attachment to their mothers; in other words, their complete "male" independence

and image are fully articulated (the opposite usually holds true of girls' initiations);

3. the dramatization of the encounter between the several generations, a dramatization that may take the form of a fight or a competition, in which the basic complementariness of various age grades —whether of a continuous or discontinuous type—is stressed; quite often the discontinuity between adolescence and adulthood is symbolically expressed, as in the symbolic death of the adolescents as children and their rebirth as adults.

4. the transmission of the tribal lore with its instructions about proper behavior, both through formalized teaching and through various ritual activities; this transmission is combined with:

5. a relaxation of the concrete control of the adults over the erstwhile adolescents and its substitution by self-control and adult responsibility.

Most of these dramatic elements can also be found, although in somewhat more diluted forms, in various traditional folk festivals in peasant communities, especially those such as rural carnivals in which youth and marriage are emphasized. In an even more diluted form, these elements may be found in various spontaneous initiation ceremonies of the fraternities and youth groups in modern societies.[4] Here, however, the full dramatic articulation of these elements is lacking, and their configuration and organization assume different forms.

The transition from childhood and adolescence to adulthood, the development of personal identity, psychological autonomy and self-regulation, the attempt to link personal temporal transition to general cultural images and to cosmic rhythms, and to link psychological maturity to the emulation of definite role models—these constitute the basic elements of any archetypal image of youth. However, the ways in which these various elements become crystallized in concrete configurations differ greatly from society to society and within sectors of the same society. The full dramatic articulation of these elements in the *rites de passage* of primitive societies constitutes only one—perhaps the most extreme and articulate but certainly not the only—configuration of these archetypal elements of youth.

In order to understand other types of such configurations, it is necessary to analyze some conditions that influence their development. Perhaps the best starting point is the nature of the social organization of the period of adolescence: the process of transition from childhood to adulthood, the social context in which the process of growing up is shaped and structured. There are two major criteria that shape the social organization of the period of youth. One is the extent to which age in general and youth in particular form a criterion for the allocation of roles in a society, whether in politics, in economic or cultural activity—aside from the family, of course, in which they always serve as such a criterion. The second is the extent to which any society develops specific age groups, specific corporate organizations, composed of members of the same "age," such as youth movements or old men's clubs. If roles are allocated in a society according to age, this greatly influences the extent to which age constitutes a component of a person's identity. In such cases, youth becomes a definite and meaningful phase of transition in an individual's progress through life, and his budding self-identity acquires content and a relation to role models and cultural values. No less important to the concrete development of identity is the extent to which it is influenced, either by the common participation of different generations in the same group as in the family, or conversely by the organization of members of the same age groups into specific, distinct groups.

The importance of age as a criterion for allocating roles in a society is closely related to several major aspects of social organization and cultural orientation. The first aspect is the relative complexity of the division of labor. In general, the simpler the organization of the society, the more influential age will be as a criterion for allocating roles. Therefore, in primitive or traditional societies (or in the more primitive and traditional sectors of developed societies) age and seniority constitute basic criteria for allocating social, economic, and political roles.

The second aspect consists of the major value orientations and symbols of a society, especially the extent to which they emphasize certain general orientations, qualities, or types of activity (such as physical vigor, the maintenance of cultural tradition, the achievement and maintenance of supernatural prowess) which can be defined in terms of broad human qualities and which become expressed and symbolized in specific ages.

The emphasis on any particular age as a criterion for the allocation of roles is largely related to the concrete application of the major value orientations in a society. For instance, we find that those primitive societies in which military values and orientations prevail

29

emphasize young adulthood as the most important age, while those in which sedentary activities prevail emphasize older age. Similarly, within some traditional societies, a particular period such as old age may be emphasized if it is seen as the most appropriate one for expressing major cultural values and symbols—for instance, the upholding of a given cultural tradition.

The social and cultural conditions that determine the extent to which specific age groups and youth groups develop differ from the conditions that determine the extent to which age serves as a criterion for the allocation of roles. At the same time, the two kinds of conditions may be closely related, as we shall see. Age groups in general and youth groups in particular tend to arise in those societies in which the family or kinship unit cannot ensure (it may even impede) the attainment of full social status on the part of its members. These conditions appear especially (although not uniquely[5]) in societies in which family or kinship groups do not constitute the basic unit of the social division of labor. Several features characterize such societies. First, the membership in the total society (citizenship) is not defined in terms of belonging to any such family, kinship group, or estate, nor is it mediated by such a group.

Second, in these societies the major political, economic, social, and religious functions are performed not by family or kinship units but rather by various specialized groups (political parties, occupational associations, etc.), which individuals may join irrespective of their family, kinship, or caste. In these societies, therefore, the major roles that adults are expected to perform in the wider society differ in orientation from those of the family or kinship group. The children's identification and close interaction with family members of other ages does not assure the attainment of full self-identity and social maturity on the part of the children. In these cases, there arises a tendency for peer groups to form, especially youth groups; these can serve as a transitory phase between the world of childhood and the adult world.

This type of the social division of labor is found in varying degrees in different societies, primitive, historical, or modern. In several primitive tribes such a division of labor has existed,[6] for example, in Africa, among the chiefless (segmentary) tribes of Nandi, Masai, or Kipigis, in the village communities of Yako and Ibo, or in more centralized kingdoms of the Zulu and Swazi, and among some of the

Indian tribes of the Plains, as well as among some South American and Indian tribes.

Such a division of labor likewise existed to some extent in several historical societies (especially in city states such as Athens or Rome), although most great historical civilizations were characterized mainly by a more hierarchical and ascriptive system of the division of labor, in which there were greater continuity and harmony between the family and kinship groups and the broader institutional contexts. The fullest development of this type of the social division of labor, however, is to be found in modern industrial societies. Their inclusive membership is usually based on the universal criterion of citizenship and is not conditioned by membership in any kinship group. In these societies the family does not constitute a basic unit of the division of labor, especially not in production and distribution, and even in the sphere of consumption its functions become more limited. Occupations are not transmitted through heredity. Similarly, the family or kinship group does not constitute a basic unit of political or ritual activities. Moreover, the general scope of the activities of the family has been continuously diminishing, while various specialized agencies tend to take over its functions in the fields of education and recreation.

To be sure, the extent to which the family is diminishing in modern societies is often exaggerated. In many social spheres (neighborhood, friendship, informal association, some class relations, community relations), family, kinship, and status are still very influential. But the scope of these relations is more limited in modern societies than in many others, even if the prevalent myth of the disappearance of the family has long since been exploded. The major social developments of the nineteenth century (the establishment of national states, the progress of the industrial revolution, the great waves of intercontinental migrations) have greatly contributed to this diminution of scope, and especially in the first phase of modernization there has been a growing discontinuity between the life of the children, whether in the family or the traditional school and in the social world with its new and enlarged perspectives.

Youth groups tend to develop in all societies in which such a division of labor exists. Youth's tendency to coalesce in such groups is rooted in the fact that participation in the family became insufficient for developing full identity or full social maturity, and that the roles learned in the family did not constitute an adequate basis for de-

31

veloping such identity and participation. In the youth groups the adolescent seeks some framework for the development and crystallization of his identity, for the attainment of personal autonomy, and for his effective transition into the adult world.

Various types of youth organizations always tend to appear with the transition from traditional or feudal societies to modern societies, along with the intensified processes of change, especially in periods of rapid mobility, migration, urbanization, and industrialization. This is true of all European societies, and also of non-Western societies. The impact of Western civilization on primitive and historical-traditional peoples is usually connected with the disruption of family life, but beyond this it also involves a change in the mutual evaluation of the different generations. The younger generation usually begin to seek a new self-identification, and one phase or another this search is expressed in ideological conflict with the older.

Most of the nationalistic movements in the Middle East, Asia, and Africa have consisted of young people, students, or officers who rebelled against their elders and the traditional familistic setting with its stress on the latters' authority. At the same time there usually has developed a specific youth consciousness and ideology that intensifies the nationalistic movement to "rejuvenate" the country.

The emergence of the peer group among immigrant children is a well-known phenomenon that usually appears in the second generation. It occurs mainly because of the relative breakdown of immigrant family life in the new country. The more highly industrialized and urbanized that country (or the sector absorbing the immigrants) is, the sharper the breakdown. Hence, the family of the immigrant or second-generation child has often been an inadequate guide to the new society. The immigrant child's attainment of full identity in the new land is usually related to how much he has been able to detach himself from his older, family setting. Some of these children, therefore, have developed a strong predisposition to join various peer groups. Such an affiliation has sometimes facilitated their transition to the absorbing society by stressing the values and patterns of behavior in that society—or, on the contrary, it may express their rebellion against this society, or against their older setting.

All these modern social developments and movements have given rise to a great variety of youth groups, peer groups, youth movements, and what has been called youth culture. The types and concrete forms of such groups varies widely: spontaneous youth groups, student movements, ideological and semipolitical movements, and youth rebellions connected with the Romantic movement in Europe, and, later, with the German youth movements. The various social and national trends of the nineteenth and twentieth centuries have also given impetus to such organizations. At the same time there have appeared many adult-sponsored youth organizations and other agencies springing out of the great extension of educational institutions. In addition to providing recreative facilities, these agencies have also aimed at character molding and the instilling of civic virtues, so as to deepen social consciousness and widen the social and cultural horizon. The chief examples are the YMCA, the Youth Brigades organized in England by William Smith, the Boy Scouts, the Jousters in France, and the many kinds of community organizations, hostels, summer camps, or vocational guidance centers.

Thus we see that there are many parallels between primitive and historical societies and modern societies with regard to the conditions under which the various constellations of youth groups, youth activities, and youth images have developed. But these parallels are only partial. Despite certain similarities, the specific configurations of the basic archetypal elements of the youth image in modern societies differ greatly from those of primitive and traditional societies. The most important differences are rooted in the fact that in the modern, the development of specific youth organizations is paradoxically connected with the weakening of the importance of age in general and youth in particular as definite criteria for the allocation of roles in society.

As we have already said, the extent to which major occupational, cultural, or political roles are allocated today according to the explicit criterion of age is very small. Most such roles are achieved according to wealth, acquired skills, specialization, and knowledge. Family background may be of great importance for the acquisition of these attributes, but very few positions are directly given people by virtue of their family standing. Yet this very weakening of the importance of age is always connected with intensive developments of youth groups and movements. This fact has several interesting repercussions on the organization and structure of such groups. In primitive and traditional societies, youth groups are usually part of a wider organization of age groups that covers a very long period of life, from childhood to late adulthood and even old age. To be sure, it is during youth that most of the dramatic elements of the transition from one age to another are manifest, but this stage constitutes only part of a longer series of continuous, well-defined stages.

From this point of view, primitive or traditional societies do not differ greatly from those in which the transition from youth to adulthood is not organized in specific age groups but is largely effected within the fold of the family and kinship groups. In both primitive and traditional societies we observe a close and comprehensive linkage between personal temporal transition and societal or cosmic time, a linkage most fully expressed in the *rites de passage*. Consequently, the transition from childhood to adulthood in all such societies is given full meaning in terms of ultimate cultural values and symbols borne or symbolized by various adult role models.

In modern societies the above picture greatly changes. The youth group, whatever its composition or organization, usually stands alone. It does not constitute a part of a fully institutionalized and organized series of age groups. It is true that in many of the more traditional sectors of modern societies the more primitive or traditional archetypes of youth still prevail. Moreover, in many modern societies elements of the primitive archetypes of youth still exist. But the full articulation of these elements is lacking, and the social organization and self-expression of youth are not given full legitimation or meaning in terms of cultural values and rituals.

The close linkage between the growth of personality, psychological maturation, and definite role models derived from the adult world has become greatly weakened. Hence the very coalescence of youth into special groups only tends to emphasize their problematic, uncertain standing from the point of view of cultural values and symbols. This has created a new constellation of the basic archetypal elements of youth. This new constellation can most clearly be seen in what has been called the emergence of the problems and stresses of adolescence in modern societies. While some of these stresses are necessarily common to adolescence in all societies, they become especially acute in modern societies.

Among these stresses the most important are the following: first, the bodily development of the adolescent constitutes a constant problem to him (or her). Since social maturity usually lags behind biological maturity, the bodily changes of puberty are not usually given a full cultural, normative meaning, and their evaluation is one of the adolescent's main concerns. The difficulty inherent in attaining legitimate sexual outlets and relations at this period of growth makes these problems even more acute. Second, the adolescent's orientation toward the main values of his society is also beset with difficulties. Owing to the long period of preparation and the relative segregation of the children's world from that of the adults, the main values of the society are necessarily presented to the child and adolescent in a highly selective way, with a strong idealistic emphasis. The relative unreality of these values as presented to the children—which at the same time are not given full ritual and symbolic expression—creates among the adolescents a great potential uncertainty and ambivalence toward the adult world.

This ambivalence is manifest, on the one hand, in a striving to communicate with the adult world and receive its recognition; on the other hand, it appears in certain dispositions to accentuate the differences between them and the adults and to oppose the various roles allocated to them by the adults. While they orient themselves to full participation in the adult world and its values, they usually attempt also to communicate with this world in a distinct, special way.

Parallel developments are to be found in the ideologies of modern youth groups. Most of these tend to create an ideology that emphasizes the discontinuity between youth and adulthood and the uniqueness of the youth period as the purest embodiment of ultimate social and cultural values. Although the explicitness of this ideology varies in extent from one sector of modern society to another, its basic elements are prevalent in almost all modern youth groups.

These processes have been necessarily accentuated in modern societies by the specific developments in cultural orientations in general and in the conception of time that has evolved in particular. The major social developments in modern societies have weakened the importance of broad cultural qualities as criteria for the allocation of roles. Similarly, important changes in the conception of time that is prevalent in modern societies have occurred. Primordial (cosmic-mythical, cyclical, or apocalyptical) conceptions of time have become greatly weakened, especially in their bearing on daily activities. The mechanical conception of time of modern technology has become much more prevalent. Of necessity this has greatly weakened the possibility of the direct ritual links between personal temporal changes and cosmic or societal progression. Therefore, the exploration of the actual meaning of major cultural values in their relation to the reality of the social world becomes one of the adolescent's main problems. This exploration may lead in many directions—

cynicism, idealistic youth rebellion, deviant ideology and behavior, or a gradual development of a balanced identity.

Thus we see how all these developments in modern societies have created a new constellation of the basic archetypal elements of youth and the youth image. The two main characteristics of this constellation are the weakened possibility of directly linking the development of personality and the personal temporal transition with cosmic and societal time, on the one hand, and with the clear role models derived from the adult world, on the other.

In terms of personality development, this situation has created a great potential insecurity and the possible lack of a clear definition of personal identity. Yet it has also created the possibility of greater personal autonomy and flexibility in the choice of roles and the commitment to different values and symbols. In general, the individual, in his search for the meaning of his personal transition, has been thrown much more on his own powers.

These processes have provided the framework within which the various attempts to forge youth's identity and activities—both on the part of youth itself and on the part of various educational agencies— have developed. These attempts may take several directions. Youth's own activities and attempts at self-expression may first develop in the direction of considerable autonomy in the choice of roles and in commitment to various values. Conversely, they may develop in the direction of a more complete, fully organized and closed ideology connected with a small extent of personal autonomy. Second, these attempts may differ greatly in their emphasis on the direct linkage of cultural values to a specific social group and their view of these groups as the main bearers of such values.

In a parallel sense, attempts have been made on the part of various educational agencies to create new types of youth organizations within which youth can forge its identity and become linked to adult society. The purpose of such attempts has been two-fold: to provide youth with opportunities to develop a reasonably autonomous personality and a differentiated field of activity; and to encompass youth fully within well-organized groups set up by adult society and to provide them with full, unequivocal role models and symbols of identification. The interaction between these different tendencies of youth and the attempts of adult society to provide various frameworks for youth activities has given rise to the major types of youth organizations, movements, and ideologies manifested in modern societies.

These various trends and tendencies have created a situation in which, so far as we can ascertain, the number of casualties among youth has become very great—probably relatively much greater than in other types of societies. Youth's search for identity, for finding some place of its own in society, and its potential difficulties in coping with the attainment of such identity have given rise to the magnified extent of the casualties observed in the numerous youth delinquents of varying types. These failures, however, are not the only major youth developments in modern societies, although their relatively greater number is endemic in modern conditions. Much more extensive are the more positive attempts of youth to forge its own identity, to find some meaningful way of defining its place in the social and cultural context and of connecting social and political values with personal development in a coherent and significant manner.

The best example in our times of the extreme upsurge of specific youth consciousness is seen in the various revolutionary youth movements. They range from the autonomous free German youth movements to the less spectacular youth movements in Central Europe and also to some extent to the specific youth culture of various more flexible youth groups. Here the attempt has been made to overcome the dislocation between personal transition and societal and cultural time. It is in these movements that the social dynamics of modern youth has found its fullest expression. It is in them that dreams of a new life, a new society, freedom and spontaneity, a new humanity and aspirations to social and cultural change have found utterance. It is in these youth movements that the forging of youth's new social identity has become closely connected with the development of new symbols of collective identity or new social-cultural symbols and meanings.

These movements have aimed at changing many aspects of the social and cultural life of their respective societies. They have depicted the present in a rather shabby form; they have dubbed it with adjectives of materialism, restriction, exploitation, lack of opportunity for self-fulfillment and creativity. At the same time they have held out hope for the future—seemingly, the not very far off future —when both self-fulfillment and collective fulfillment can be achieved and the materialistic civilization of the adult world can be shaken off. They have tried to appeal to youth to forge its own self-identity in terms of these new collective symbols, and this is why they have been so attractive to youth, for whom they have provided a set of symbols, hopes, and aims to which to direct its activities.

Within these movements the emphasis has been on a given social group or collectivity—nation, class, or the youth group itself—as the main, almost exclusive bearer of the "good" cultural value and symbols. Indeed, youth has at times been upheld as the sole and pure bearer of cultural values and social creativity. Through its association with these movements, youth has also been able to connect its aspiration for a different personal future, its anxiety to escape the present through plans and hopes for a different future within its cultural or social setting.

These various manifestations have played a crucial part in the emergence of social movements and parties in modern societies. Student groups have been the nuclei of the most important nationalistic and revolutionary movements in Central and Eastern Europe, in Italy, Germany, Hungary, and Russia. They have also played a significant role in Zionism and in the various waves of immigration to Israel. Their influence has become enormous in various fields, not only political and educational but cultural in general. In a way, education itself has tended to become a social movement. Many schools and universities, many teachers, have been among the most important bearers of collective values. The very spread of education is often seen as a means by which a new epoch might be ushered in.

The search for some connection between the personal situation of youth and social-cultural values has also stimulated the looser youth groups in modern societies, especially in the United States, and to some extent in Europe as well—though here the psychological meaning of the search is somewhat different. The looser youth groups have often shared some of the characteristics of the more defined youth movements, and they too have developed an emphasis on the attainment of social and cultural change. The yearning for a different personal future has likewise become connected with aspirations for changing the cultural setting, but not necessarily through a direct political or organized expression. They are principally important as a strong link with various collective, artistic, and literary aspirations aimed at changing social and cultural life. As such they are affiliated with various cultural values and symbols, not with any exclusive social groups. Thus they have necessarily developed a much greater freedom in choice of roles and commitment to values.

Specific social conditions surround the emergence of all these youth groups. In general, they are associated with a breakdown of traditional settings, the onset of modernization, urbanization, secularization, and industrialization. The less organized, more spontaneous types of youth organization and the more flexible kind of youth consciousness arise when the transition has been relatively smooth and gradual, especially in societies whose basic collective identity and political framework evince a large degree of continuity and a slow process of modernization. On the other hand, the more intensive types of youth movements tend to develop in those societies and periods in which the onset of modernization is connected with great upheavals and sharp cleavages in the social structure and the structure of authority and with the breakdown of symbols of collective identity.

In the latter situation the adult society has made many efforts to organize youth in what may be called totalistic organizations, in which clear role models and values might be set before youth and in which the extent of choice allowed youth is very limited and the manifestations of personal spontaneity and autonomy are restricted. Both types of conditions appeared in various European societies and in the United States in the nineteenth and early twentieth centuries, and in Asian and African societies in the second half of the twentieth century. The relative predominance of each of these conditions varies in different periods in these societies. However, with the progress of modernization and the growing absorption of broad masses within the framework of society, the whole basic setting of youth in modern society has changed—and it is this new framework that is predominant today and in which contemporary youth problems are shaped and played out.

The change this new framework represents is to some extent common both to the fully organized totalistic youth movements and to the looser youth groups. It is connected mainly with the institutionalizing of the aims and values toward the realization of which these movements were oriented, with the acceptance of such youth organizations as part of the structure of the general educational and cultural structure of their societies.

In Russia youth movements became fully institutionalized through the organization of the Komsomol. In many European countries the institutionalizing of youth groups, agencies, and ideologies came through association with political parties, or through acceptance as part of the educational system—an acceptance that sometimes entailed supervision by the official authorities. In the United States, many (such as the Boy Scouts) have become an accepted part of community life and to some extent a symbol of differential social status. In many Asian and African countries, organized youth movements have become part of the nationalistic movements and, independence won, have become part of the official educational organizations.

This institutionalizing of the values of youth movements in education and community life has been part of a wider process of institutionalizing various collective values. In some countries this has come about through revolution; in others, as a result of a long process of political and social evolution.

From the point of view of our analysis, these processes have had several important results. They have introduced a new element into the configuration of the basic archetypal elements of youth. The possibility of linking personal transition both to social groups and to cultural values—so strongly emphasized in the youth movements and noticeable to some extent even in the looser youth culture—has become greatly weakened. The social and sometimes even the cultural dimension of the future may thus become flattened and emptied. The various collective values become transformed. Instead of being remote goals resplendent with romantic dreams, they have become mundane objectives of the present, with its shabby details of daily politics and administration. More often than not they are intimately connected with the processes of bureaucratization.

All these mutations are associated with a notable decline in ideology and in preoccupation with ideology among many of the groups and strata in modern societies, with a general flattening of political-ideological motives and a growing apathy to them. This decline in turn is connected with what has been called the spiritual or cultural shallowness of the new social and economic benefits accruing from the welfare state—an emptiness illustrated by the fact that all these benefits are in the nature of things administered not by spiritual or social leaders but, as Stephen Toulmin has wittily pointed out, "the assistant postmaster." As a consequence, we observe the emptiness and meaninglessness of social relations, so often described by critics of the age of consumption and mass society.

In general, these developments have brought about the flattening of the image of the societal future and have deprived it of its allure. Between present and future there is no ideological discontinuity.

The present has become the more important, if not the more meaningful, because the future has lost its characteristic as a dimension different from the present. Out of these conditions has grown what Riesman has called the cult of immediacy. Youth has been robbed, therefore, of the full experience of the dramatic transition from adolescence to adulthood and of the dramatization of the difference between present and future. Their own changing personal future has become dissociated from any changes in the shape of their societies or in cultural activities and values.

Paradoxically enough, these developments have often been connected with a strong adulation of youth—an adulation, however, which was in a way purely instrumental. The necessity of a continuous adjustment to new changing conditions has emphasized the potential value of youth as the bearers of continuous innovation, of noncommitment to any specific conditions and values. But such an emphasis is often couched in terms of a purely instrumental adaptability, beyond which there is only the relative emptiness of the meaningless passage of time—of aging.[7]

Yet the impact on youth of what has been called postindustrial society need not result in such an emptiness and shallowness, although in recent literature these effects appear large indeed. It is as yet too early to make a full and adequate analysis of all these impacts. But it should be emphasized that the changes we have described, together with growing abundance and continuous technological change, have necessarily heightened the possibility of greater personal autonomy and cultural creativity and of the formation of the bases of such autonomy and of a flexible yet stable identity during the period of youth.

These new conditions have enhanced the possibility of flexibility in linking cultural values to social reality; they have enhanced the scope of personal and cultural creativity and the development of different personal culture. They have created the possibility of youth's developing what may be called a nonideological, direct identification with moral values, an awareness of the predicaments of moral choice that exist in any given situation, and individual responsibility for such choices—a responsibility that cannot be shed by relying on overarching ideological solutions oriented to the future.

These new social conditions exist in most industrial and postindustrial societies, sometimes together with the older conditions that gave rise to the more intensive types of youth movements. They constitute the framework within which the new configuration of the archetypal elements of youth and the new possibilities and problems facing youth in contemporary society develop. It is as yet too early to specify all these new possibilities and trends: here we have attempted to indicate some of their general contours.

REFERENCES

1. A general sociological analysis of the place of age in social structure has been attempted in S. N. Eisenstadt, *From Generation to Generation* (Chicago: The Free Press of Glencoe, Illinois, 1956).

2. The analysis of personal, cosmic, and societal time (or temporal progression) has constituted a fascinating but not easily dealt with focus of analysis. For some approaches to these problems, see *Man and Time* (papers from the Eranos Yearbooks, edited by Joseph Campbell; London: Routledge & Kegan Paul, 1958), especially the article by Gerardus van der Leeuw. See also Mircea Eliade, *The Myth of the Eternal Return*. Translated by W. R. Trask. New York: Pantheon Books, 1954 (Bollingen Series).

3. For a fuller exposition of the sociological significance of initiation rites, see Mircea Eliade, *Birth and Rebirth* (New York: Harper & Brothers, 1958) and *From Generaton to Generation* (ref. 1).

4. See Bruno Bettelheim, *Symbolic Wounds, Puberty Rites and the Envious Circle* (Chicago: The Free Press of Glencoe, Illinois, 1954).

5. A special type of age groups may also develop in familistic societies. See *From Generation to Generation* (ref. 1), ch. 5.

6. For fuller details, see *From Generation to Generation*, especially chs. 3 and 4.

7. For an exposition of this view, see Paul Goodman, "Youth in Organized Society," *Commentary*, February 1960, pp. 95-107; and M. R. Stein, *The Eclipse of Community* (Princeton: Princeton University Press, 1960), especially pp. 215 ff.; also, the review of this book by H. Rosenberg, "Community, Values, Comedy," *Commentary*, August 1960, pp. 150-157.

KASPAR D. NAEGELE

Youth and Society

Some Observations

HERODOTUS REPORTS a story:[1]

Two youths, Kleobis and Biton, were both distinguished for their re-markable strength, and had won many a victory in the gymnastic games. Because oxen were missing, they pulled their mother, a priestess of Hera, in her chariot a great distance to the sanctuary at Argos. For this pious deed, their mother prayed the gods to reward them and, as the greatest boon they could grant, the gods allowed the two brothers to die in their sleep in the full strength of their youth.

It is my belief that, for all our emphasis on the qualities of youth, we today would not create the equivalent of this tale. We do not want to die in our sleep in the full strength of our youth. Yet we still share with the ancient Greeks the wish that "youth should not be spoiled by old age."[2] We try to stay young. We would not ask the ageless gods to snatch their gift, our young lives, from us and so preserve us from the earthly loss of them. Instead, we hope that we shall persist in living and achieving, and that our age will not show in the process. And our youths, like ourselves, are to prepare themselves for a fairly long life of effort and advancement.

This contrast between some Greek and American desires, as well as the coincidence of contrary themes within our own society, constitutes the subject of this essay. What I am ultimately concerned with is the relation between individual life cycles (whether they are complete circles or incomplete arcs) and changes in social patterns. More particularly, I am concerned with education as a stream of occasions in which youth is encountered and transformed, and with the fact that youth cannot last. Therefore, a society must make room for youth and provide doors for leaving it behind. My thoughts in these

43

regards are directed by a series of distinctions to be presently suggested, a set of ongoing explorations to which this essay will also refer, and the paradigmatic contrast of the Greek tale with American hopes and with the outcry of a young German named Wolfgang Borchert. Born in Hamburg, he died on 20 November 1947 in Basel, of an illness contracted from imprisonment, military service, and starvation. At his death he was twenty-six years old. He left behind him a volume of poems, short stories, essays, and a play, *Draussen vor der Tür.*

Whether the intensity of his writing springs from his youth or is the expression of a condensed life, I do not know, nor is it clear how such a man might have developed had medicine been able to save him. One might argue that in his desolate descriptions of a youth in a defeated country, at the end of a dozen years of Nazidom and war, he speaks effectively but only for his own moment in time. Perhaps. In one short essay, however, he writes eloquently about his image of his generation—a generation *ohne Abschied.* Let me quote the beginning and concluding sentences:[3]

Wir sind die Generation ohne Bindung und ohne Tiefe. Unsere Tiefe ist Abgrund. Wir sind die Generation ohne Glück, ohne Heimat und ohne Abschied. Unsere Sonne ist schmal, unsere Liebe grausam und unsere Jugend ist ohne Jugend. . . . Aber wir sind eine Generation der Ankunft. Vielleicht sind wir eine Generation voller Ankunft auf einen neuen Stern, in einem neuen Leben. . . . Wir sind eine Generation ohne Abschied, aber wir wissen, dass alle Ankunft uns gehört.

To begin with, I cannot help viewing youth as the shrouded inheritance of childhood; but this immediately introduces the distinction between a view from within and one from without. Presumably, this distinction depends on age. It demands that we ask how youths themselves think and feel and how their judgments differ, both from those who belong to other generations and from those who seek primarily to be observers, probing the discrepancy (and coincidence) between ascertainable fact and guiding belief. The very expression "youth," in other words, reminds us of a universal pluralism of thought and emotion—the pluralism inherent in the life cycle. Quite apart from their bodily appearances, we expect the difference between children and youths, between youths and their elders, and between older people and their grandchildren to lie in the way they think. We may assume this includes both the range of the matters they ponder and the style in which they ponder them. Time, for instance,[4] is different for young adults and for older persons. The old may do less, but time for them goes by faster; at least, they prefer to describe time by metaphors that suggest swiftness: a galloping horseman, a fleeing thief, a fast-moving shuttle. Young people, for all their greater involvement, are more prepared to think of time as a large and static quantity: a road leading over a hill, a quiet ocean, the Rock of Gibraltar. For the young, there is indeed time; more is yet to come than has already gone by. For the old, time is swift because it is scarce, for more time has gone than is left.

If societies are alike in sustaining a contrast in perspective between people of different ages, they surely differ in the constancy of this contrast. If societies divide the interval between birth and death —embedding it perhaps in a wider circle of continuous or immortal life, or seeing it as a self-contained appearance—into stages of a life cycle, or of a career, then they surely differ in the extent to which the experience of the present generation is a guide to the future or the past. Here (to anticipate) our youths surely stand in a novel position. As citizens in a society that is in continuous transition, they can be sure that their fathers' youth was substantially different from their own and that their children's youth will again be different. In addition, the various views from "within" will be different from the views from "without." The grown adult, not only the youth and the child, may confront a poignant discrepancy between what he believes and what someone else can show him to be fact, or between what he knows and what (had he more time) he could know.

It follows, then, that youth, as a stage in a shifting succession of stages, is nowadays bound to be seen from both without and within. Its special character perhaps lies precisely in the fact that from now on the several views of differing people may shape the inner life of a single person. Thus in youth the irreducible facts of aloneness are established.

Youth as a stage raises the further distinction between the actual qualities of specific young people and the qualities that together constitute people's different conceptions of youth. I do not know whether all societies think of certain years in the life cycle as "the best." With us, in contemporary North America, these tend to be the years of our youth, when we are first married and have not as yet the responsibility of caring for young children of our own. So, at least, our norms would have us believe. The facts may be quite different. Thus it comes about that youth is at once a stage of life and also a form of living. In our case, it stands presumably for a characteristic mixture of enthusiasm, intensity, and eagerness. It stands for anticipation—

the anticipation sustained by the will to try, not by sheer wonderment at the nature of things. It stands for being active but not scheming. Actually, we know that these qualities, though considered indicative (if not constitutive) of youth, may in fact be absent, though flourishing among "older" people, who thus seem younger. In that sense, youth as a form may seem absent from the lives of some and permanently embedded in the lives of others.

Finally, both as a stage and as a form, youth drives a wedge into the continuity of the cycle of life. It shares with other stages and forms the ability to give life a "severalness" that goes beyond the alternation of wakefulness and sleep or of hunger and satiety. An adventure (or a misadventure) as a form of heightened living stands in contrast to routine: it interrupts or disrupts, and therefore helps give routine its form, its direction, and its end. Youth also disrupts: it drives a wedge between childhood and adulthood. The triangle thus made, of course, contains a sequence of finer and smaller transitions. Besides, life stretches beyond the boundaries of this triangle as adulthood, itself no monolithic period of time, becomes transformed into old age. But the wedge of youth results precisely from the fact that youth must seek—and suffer—a discontinuity from childhood: it must be proud of leaving childhood behind in order to qualify for adulthood.

Much is expressed in these acts of qualifying. As one increasingly engages in such acts, childhood is irrevocably left behind. To qualify usually means to acquire a specific or a general competence which another person (already qualified) can now recognize as a right to assume new and wider privileges and obligations. Youth claims a new autonomy or a new membership. Adults, who by definition are endowed with more power or more honor, may allow, withhold, or limit this claim. Thus youth's claim to be recognized in its demands or its rights is at once a recognition of its dependence on adults, whose views still matter, and a bid for independence. In extreme anger or despair, youth may also seek to bypass independence by pitting its self-conception against the appraisal of its elders. Youth then tries to displace such an appraisal by withdrawing into activities of its own, insulated from the affairs of the previous generation. The practical limits to such rebellion and withdrawal sometimes strengthen further the desire for them. In this fashion, utopian and youthful dispositions thrive on one another. As a category of age, youth then wears a poignantly double face.

Suspended between a "no longer" and a "not yet," youth is forced to balance continuity and discontinuity. To a degree, this may be true of all ages in our era. But children have less to look back on, less to leave behind, as they move into the future. Adults have fewer radically new experiences to move into as they leave the past. To youth, the present contrasts much more vividly with past and future. This contrast, as we know, is composed of diverse elements: inner processes of growth and rebellion, surrounding images of the qualities of youth in the form of ideals and judgments sustained by people occupying different spaces in individual and collective time-tables, and patterned opportunities for living—whether licit or illicit, or occasions of learning, working, playing, or, in the wider sense, loving. Youth, under our circumstances and in the context of our liking for vigor, innovation, and improvement is likely to harden its contrast to childhood and adulthood into an ideal or actual pattern of "self-containedness," an island of heightened sensibility. From there childhood looks long ago, and the promised land is only partly visible in the shape of contemporary adults.

It is not possible here to be systematically explicit either about the major distinctions that constitute my mode of thought or about the array of specific propositions which eventually I should like to revise or confirm. This much, however, needs saying.

1. In our society, youth follows a childhood much of which is spent in a formal school system which balances the right to an education with the obligation, for parents and children, to avail themselves of it. This school system, too, is divided into certain main forms (public, private, and parochial) that in turn become embodied in the self-images of the pupils. Moreover, by the time the latter are fourteen, they will have had to learn to spend a good part of the day for a good part of the year in the continuous presence of their contemporaries, faced by a rather special selection of adults, their teachers. They will have been asked to devote themselves rather single-mindedly to the enterprise of learning and being tested. Earning a living—and living—will often appear apart from the school and from home: a promised and dreaded territory elsewhere in space and time.

2. Youth in industrial societies is exempt or kept from most of those positions and opportunities which allow a fairly continuous involvement in decisions of a political, economic, or military character that affect large numbers of people. It wields relatively little formal political authority.

3. In an industrial and modern society, adults interested in en-

hancing their independence or their prominence have certain alternatives. Yet, whether they seek to be creative in the arts, or content at home, or powerful in public, they will be expected to recognize this world for what it is. If they do not, they will be considered naive—a description that explicitly recognizes that adults are expected to have a knowledge of the impersonal character of many important human arrangements and of the informal labyrinth that helps make a formal framework effective. In the same sense, we would not expect youth to be enthusiastically knowledgeable about legal, economic, or bureaucratic patterns, for these would seem dull and slow (and perhaps corrupt) compared with a youth's concern with establishing himself or herself. Yet this concern, oscillating between intense hope and despair, can also evoke a clear-sighted if not tempered recognition of the "deals" by which the world carries forward in practice the competing ideals that, often in disembodied form, adults propose to children. Under these conditions, youth seems to have an especially long way to go in an industrial society in which adults must so frequently maintain public life by reasoning, planning, calculating, or scheming. In contrast, youth still stands for spontaneous, free, and unself-conscious activities, for a willingness to express enthusiasms as well as to have them.

4. Contemporary youth in the United States lives in a society that as a whole considers itself young, at least in contrast to its European ancestors. The image of the youth of America is now, of course, complicated by the attention given underdeveloped societies. (To provide these societies with counsel and aid, when these are sustained, is presumably one further mark of the end of youth.) American society, therefore, finds itself collectively constrained to advance its self-image from one of youth to a less chronologically specific and more demanding conception of a "large and powerful nation," while finding its young people rather problematic. While the country is indeed demographically aging, it has to pay special attention to education, an activity normally associated with youth.

As modern men or as scholars, we can hardly claim that the historic transformations that link contemporary North American society to other and past societies are closely marked by progress or even by some less evaluative and more specific direction. We might agree that there is enough evidence for us to speak about the rationalization of the world. Also, within the last three hundred years, in the Western world, we have progressively freed childhood from the grosser cruelties of physical punishment, hard labor, and punitive teaching. We are more prepared than in previous centuries not to think of children as small-scale editions of adults, but to think of adults as large-scale editions of children. Within this perspective, we are further prepared to think of childhood as a period *sui generis,* as well as a determinative beginning for subsequent possibilities. We may well have substituted subtle forms of anxiety and uncertainty for the grosser forms of neglect or cruelty that we recognize as having existed in the past. Nor is adolescent life in the twentieth century part of this striving for greater care of and concern for children. Besides, comparatively speaking, American society, in contrast to primitive cultures, is not more loving, certain, and liberating in its patterned ways of attending to children and enabling youth to proceed into adulthood.

Rather, we have become convinced that childhood casts its lights and shadows on what follows it and that later demands or possibilities require some and rule out other forms of earlier experience. We want to spare our children certain fears, and we expect of ourselves that we should understand them, so that they, feeling understood, will in turn have reason to enjoy us. In practice, of course, the situation is considerably more complicated and varied. Nor are the consequences of our acts necessarily in line with our intentions. Childhood, however, has become the object of self-conscious discussion and deliberate professional intervention on the part of those to whom previous traditions seem insufficient.

This digression is sustained by value judgments, including the belief that ultimately human arrangements can neither be lived nor studied without some commitment to ideas and ideals which define and refine the several domains of play, work, and love, both in themselves and in their mutual enhancement and limitation. In this connection I have been struck by certain impressions, none of which have I had the opportunity or skill to discipline into propositions that can be proved or disproved.

1. As I look at portraits of children of the nineteenth or eighteenth century, their faces look both sadder and older than their contemporaries' would today.[5] Painting is not photography, of course, and both are selective. Turn to the Renaissance or the early Middle Ages, and the picture becomes even more complicated. Still, we have increasingly come to expect childhood to be in some sense happy, light, carefree. Yet, as childhood moves more and more into the province of the schools, we come to have various second thoughts as

49

to this apparent lightheartedness. The debate about education is directly a debate about priorities and realities in adult life; it is also a debate whose alternative solutions affect the scope and constraint we give to children.

2. There has been a continuous debate about education, a continuing shift in the status of teachers, and yet the influence that teachers have on the course of youth is not yet adequately understood. We may exaggerate or underestimate it. We have yet to discover, however, what teachers mean when society itself changes radically, as it did in Germany in 1933. Then youth can compare a "before" and "after" and can learn, perhaps painfully, to distinguish between inner growth and the kind of personal change which is only an expedient response.

3. In the free societies of today, youth is surrounded by elders who generally feel rather helpless concerning the course of events for youth. I believe this has stimulated a certain kind of withdrawal into private life, a withdrawal which is consonant with some of the utopian elements within the pattern of dominant values in the society of our youths' elders.

Even a rudimentary sense of the history and variety of human society leaves the impression, confirmed by further probing, that the experience of being young, or even just of being a youth, yields rich diversities, whether one considers primarily an inner world or also the kinship ties, school systems, age groups, or other youth-affecting social patterns that engender, limit, or direct the inner world of youth.

Within the mainstream of industrializing societies in the West (including Russia and beginning with their antecedents, notably classical Greece), there has been a rich variety of educational enterprises. One can discern their structure and their sponsorship, the qualities they hope to engender, the strengths that sustain them in supportive or abortive environments, or the precise contributions they make to the inner life or visible accomplishments of their pupils.

At first, the variety seems endless, if not random. Think of the academies of classical Greece, of the contrasts between Athens and Sparta. Think of Italy during the Renaissance, where Vittorino de Feltre educated the sons and daughters of aristocratic families in Mantua and taught them the new humanism and its version of rhetoric and science, while also instructing them in the proper development of their muscles. Nor was he concerned merely with the children of the privileged. He cared also for the children of talent and the children of the poor. In his house they all lived together, their diversity constrained within a religious boundary.[6]

Think of the Danish folk high schools that began to flourish in the nineteenth century, drawing inspiration from a mixture of religious and patriotic motives. Grundtvig[7] and his followers had fairly definite ideas about the timing of education. They wanted childhood and early youth to be free. They thought of intellectual discipline as an imposition that should come only near adulthood. They wanted in general, especially for rural Danes, an education for life that would strengthen a self-assured and nonaggressive cultural autonomy vis-à-vis the neighboring Germans. Meanwhile, the Germans further differentiated a more limited, earlier ideal as embodied in the *Gymnasium;* they added various *Realschulen,* and sought to make education more flexible, open, and inclusive. They tried to combine humanistic and technical interests and so to reconcile the ideals of the cultivated man (in his Greek version rather than in the form of the English gentleman) with the culturally more transcendent and qualitatively emptier ideal of the competent specialist. At least one scholar[8] argues at length in one of those circumspect yet probing works of the nineteenth century which combine a wide view with detail and certainty, that such an ambition begins with *Utraquismus* and ends in *Überbürdung* for youth. As a learner and a youth, one ends, in other words, with a sense of excessive expectation, insufficient time, and the continuous demand to coordinate divergent intellectual tasks within a fairly rigid timetable. In America today we hear the reverse; that we have allowed ourselves to expect too little of our youth. In any event, youth poses the problem of expectation. Nor is this ever only a matter of quantity.

Or again, think of Sweden: a Swedish school child at the time of Gustav Adolf's accession at the end of the sixteenth century went perhaps to a *trivium* school and learned Latin. He learned at most "a bare minimum of religious knowledge," in addition to grammar, rhetoric, and dialectic. He began early in the morning and somehow sustained a fifty-one-hour school week. The dull rigors of the scheme were only slightly relieved by the chance of holidays, sometimes to collect nuts; sometimes to dance. The *trivium* school (its curriculum was modeled on the mediaeval *trivium*) was run by a poor private schoolmaster answerable only to his bishop. If he was to get additional staff, he had to pay for it himself, unless he used pupil teachers who were likely to help with the younger children. The buildings were dim and crowded, the discipline severe. In the main, the pupils came from the middle or lower classes, for the nobility had private tutors. The pupils, often as poor as their teachers, belonged like them to a community of scholars that depended on public charity. Indeed, twice a year, at Christmas and harvest time, they "sang for their supper," visited assigned parishes, and collected money and goods which in part at least came back to their schools. *Sockengang*, as this was called, allowed a young person quite a range of enterprises from teaching children, interpreting the Bible, or saying prayers over the sick (animal or human) to converting into money what others put into his leather bag. Besides, this escape from the tedium of oral recitations, chorused by some in a classroom while others studied in quiet, could be further exploited in pranks, fights, drinking, or the beginnings of a career in begging.[9]

And then think of the slow transformation and extension of the more conventional system of public education, which has bequeathed us our own system.

This essay is shaped by the quadrangle of childhood, youth, adulthood, and old age; in the foreground, moreover, lies the triangle of these first three divisions, which makes of youth a period of "outsideness." I know these are anything but universal images. Clearly, cultures vary in the time-tables they devise for the life cycle. Classical China, for instance, recognized a succession of stages which differs from our own.[10]

Nor is the number of stages in itself the important issue. Rather, the *scheme* of contrasts (and hence of continuities and discontinuities) as usually experienced or as necessary for the analysis of a person's temporal location within a life span is at issue. Erik H. Erikson for one has suggested a succession of stages. I do not know whether these are best (or uniquely) descriptive of the inner life of Western man. The succession is given impetus by the pervasive presence of organizing dilemmas. Again, is this true in terms of our experience, or is it in addition necessary for the explanation of "why we feel the way we do?" Besides, I am concerned with both individual youths and youth as a personal time span which creates social bonds and social patterns. The latter include youth cultures and youth movements.

To be sure, the existence of the latter helps direct the inner experience of one's youth, even if (as in North America today) we can elaborate or bypass youth culture and youth movements, and so add novelty to the chronically unfolding contrast of the generations. It may be best to avoid schematizing with definitions and fourfold tables, but it is necessary to distinguish at least the following: youth as an *inner condition;* youth as an *object of images* (held by those not yet, or never, or no longer belonging to youth and by those who in their own appraisal do belong); the images of youth *in the company* of other images giving character to a given society; and the *arrangements* (including special persons such as teachers) responsible for and to youth. I think of these aspects of youth as arranged in a circle, influencing one another.

As an inner condition, youth in our society today, as I hear from university students, is a period of intermission between earlier freedoms (or so they now appear) and subsequent responsibilities and commitments. It is a last hesitation before certain rather serious commitments concerning work and love move one fully into or against the wider conventional life. It is an opportunity for exploration, even though it is also a period of deprivation, the deprivation that comes from having to meet requests which are ill-fitted to one's capacities or attitudes, requests which are too much or too little, and which proceed from assumptions that seem alien. The familiar ambivalence of youth—its impulsiveness and fastidiousness, its wish for autonomy and dependence, its scepticism and intensity—must here go unexamined. The precise character and relative distribution of this ambivalence, by which we believe we recognize adolescence, remains to be properly documented and established. After all, the descriptions of youth are more often than not the claims of adults or the reflections of youth wishing to speak to adults. Many distortions thrive here. But to listen to youths speaking to one another, though very necessary, would also be insufficient. Youth surely is nothing (or everything, in the form of a utopia) if not a contrast to another, larger part of the life span.

In that sense, in our circumstances, youths balance their isolation from adults with an intense (if often negative) relation to them. This Western discontinuity is of course much transformed under totalitarian conditions, and in addition it has of late taken on the form of a kind of passive withdrawal. With us, rebellion—by which an independence is wrought and the newness of a new generation, or at least the oldness of a previous one, is proclaimed—often takes the form of a quietistic emphasis on personal experience and exploration rather than of outward political radicalism. It can, of course, also take the form of the familiar delinquencies. To the extent to which the latter involve heroin or marihuana, they come to fit the same pattern.

At least two judgments seem implied by a withdrawal from the public domain into that of personal experience: the judgment that human and private matters are more important than concern with power, money, success, or even justice; and the judgment that this society provides no scope for a satisfying enactment of one's ideals, at least for the ideals of youth. This withdrawal has been too much regarded as a search for security, and not enough as a rebellion against convention and popular values. Nor is it clear to what extent this rebellion has in turn been accompanied by the accumulation of a reservoir of unenacted idealism. The fortunes of the Peace Corps will perhaps provide some evidence in this respect. To be sure, rebellion, as one aspect of the tension between youth and the adult world, also exists in direct forms. For all its special features, the Crusaders in America are a small but important example. Student rebellions in totalitarian or dictatorial milieux (Germany, Hungary, and the like) provide further qualifications. Besides, the contrast between youth and adult society, which with us takes the form of a particular kind of distance, surely demands overriding continuities under totalitarian conditions. (I hope to return to this point later.) In our society, the distance between youths and adults is a cherished if distressing fact. It produces in youths the simultaneous impressions of thinking they are wrongly appraised, that they are happily different from what those who only partly understand them think they are, and that they are better than their elders, whose good opinion, nevertheless, they still want.

Recently I asked some two hundred university students between sixteen and twenty-two whether they would want people under twenty-one to be allowed to vote in Canadian federal elections. About half said no. On my asking a few for their reasons, the answers all converged on the opinion that before twenty-one one's judgments are both unstable and easily influenced. Is this the other side of an equation of maturity with certainty, and hence of an identification of uncertainty with youth, in whom the pains of puzzlement are balanced by the freedom from certain obligatory decisions? Or are young people willing to accept their elders' judgment that youth is indeed too young to know—or at least, to make decisions?

I also asked how they might spend a year with any amount of

money at their disposal. Again, fifty percent agreed: they would neither refuse the money nor save it, they would travel after graduation. (Only three would refuse it, while three would save the money.) When asked where they would like to live while studying (money again being no obstacle), a quarter of the group would prefer to live with their parents, the same number would prefer university residence, and about four out of ten would prefer to live alone, away from home and parents.

Let me finally return to youth as it provides opportunities for those no longer young. I was and am interested in teaching as a social role, as compared with healing and "saving" people.[11] Among other things, I therefore asked two dozen teachers (twelve of each sex) what they took to be the main differences between children and adults. Their first response was a rather stunned silence, followed by the protest that I should have warned them of such a question so that they might have prepared an answer. For my part, I had expected them to find such a question difficult.

The women teachers saw children in positive terms, in the main. More, they thought of them in comparative terms. They accepted the question, divided the life cycle into two, and proceeded to contrast adults with children. In this comparative mode, children seemed to them primarily honest. In the same breath, with more emphasis, children were exempted from hypocrisy. As one teacher put it,

. . . there is nothing like hypocrisy in them . . . they say exactly what they think, not like adults, who are inclined to beat about the bush.

Honesty and its associated qualities of openness and integrity were commended as virtues, even while some of the teachers acknowledged as well that children in their honesty can hurt the feelings of others. Therefore, one has to be direct with them and, alas, teach them to contain their frankness.

Second, however, children can be taught, they want to learn, and as a rule they have a rewarding curiosity. In these respects, too, they exceed adults. The men teachers agreed as to this, but were more evasive, being more involved in the tasks of development that childhood bestows on children and in the tasks that they as teachers had set themselves to accomplish with their pupils. I also asked both groups of teachers what they most enjoyed about children. Again they agreed on one main experience: they saw children as responsive individuals who would leave them at the end of the school year in a visibly enhanced state, having learned or mastered diverse things and remaining fully engaged in an onward development. This is not the place to amplify or question the teachers' answers, to distinguish asserted opinion from actual belief and practice, or to puzzle over the fact that children by no means say all they think, and that often they are forced to think only what they can say rather than what they deeply feel.

Yet what have the thoughts of some teachers, in British Columbia at that, to do with an interest in youth in any other part of the world, not just in the United States? The teachers, it seemed to me, were combining a reassuring delight in the growth of their pupils with a wise sadness that this cherished growth meant in fact a loss of certain cherished qualities, as well. Adults, as the end product of a process of learning to which they as teachers were called on to foster, did indeed seem inferior to children. The interviews did not allow time to explore this dilemma. Clearly, by the logic of their own accounts, the teachers had admitted that they had to help in the transformation of children into adults, and that hence they were contributing in some measure to the loss in children of qualities which in their full expression they found so satisfying.

I have discussed these interviews for two reasons. One way to understand the varieties of youth that history shows us is to examine the provisions made for educating youth. These provisions, of course, are a compound: they are informed by aims that are themselves derived from a wider pattern of beliefs, which in turn constitute the dominant commitment of a society. In that respect, education is a means of instituting a wider social order. In addition, this is an order in its own right, one that both widens and limits the expectations it gives rise to. We know how teachers as a professional group and school systems as a form of bureaucratic organization present certain issues that complicate the journey from childhood to adulthood. Yet the preferred relation of teachers to youths, together with the reciprocal relations youth has with others (including other youths) is surely one element in its experience. The relative importance of its schooling, as compared with its kinship relations, its membership in a political party or a religious group, or its participation in joint enterprises with its peers, must be left entirely open—the more so since the sector of modern industrial society in which youth is spent surely makes a difference.

My second reason for citing this interrogation of students is that I should like to characterize further this romantic image of youth, and then contrast it with alternative ones. Ultimately, this romantic view is one that equates virtue with innocence and unself-consciousness. Children are frank, open, and curious creatures, intent on a free exchange between themselves and the world, and as such they are a refreshing contrast to their more closed, more hesitant, more devious elders. Yet the latter, for all their romanticism, if they are teachers, also value knowledge and demand that it be acquired. This incongruity troubles their relations with children, and these troubles multiply when children leave childhood to enter the ranks of the elders via adolescence. The contradictions in this romantic view give rise to intensely positive and negative attitudes on the part of both youths and adults.

Furthermore, this romantic view is one of a trinity of complementary and contrary dispositions. In the West, it is the obverse side of a Christian religious tradition that would replace the innocence of childhood with the belief in original sin. In this light, youth becomes an opportunity for "confirming" the process of salvation, and the indoctrinated become incorporated into one or another religious communion. If the romantic image emphasizes the goodness of a state of nature which is lost to society (at least, in part), the Christian view emphasizes the need for the salvation of youth.

In contrast to these two opposing views, there is a third, which suggests several further alternatives. This is the secular view that ignores any simple moral qualities in childhood, regards it as "polymorph and perverse," and sees youth as a period of search and a rebellion (active or passive), which can finally constitute a person capable of work and love—and as such the progenitor of others. The Jewish and Greek traditions represent further variations in a similar direction. Both, I think, combine a genuine concern for education with significant inadequacies in providing opportunities for it. The Greek ideals of *paideia* are surely intended primarily for man: they imply the familiar classical conception of form, wisdom, and beauty. They imply a progressive self-knowledge as the necessary means for attaining wisdom, and a turning away from a youthful concern with external and natural order toward a reasoning knowledge of internal facts and the ordering of these facts by a knowing self. Youth then becomes the opportunity *par excellence* for the full institution of a pattern of education shaped by nonromantic ideas of excellence. This pattern, however, leads to a definitely stratified, intellectualized society in which doubt and argument co-rule with the definitive disclosures of mathematics. The Jewish tradition, on the other hand, which is concerned with justice and the use of the household as a religious and ritualistic group to which youth belongs from infancy, is less rationalistic than the Greek, though perhaps not less intellectual. It also regards the education of children as a serious matter; in the earliest form of Judaism, it was neither in the hands of women nor of single men: married men instructed the coming generation in the knowledge of the Law. Youth in this tradition is a valued resource and its teachers are honored, the more so since learning is conceived of as a life-long, not a youth-contained, enterprise.

Today, our resources for both enhancing and burdening youth go beyond the various combinations of romantic, Christian, Greek, and Jewish traditions, and beyond the blend of these traditions with those of the Anglo-Saxon, French, Italian, German, Scandinavian, or other Western cultures, in at least two major respects: modern industrial societies, regardless of their political commitment, seem to produce a special configuration for youth; and within this configuration, totalitarian versions of industrial societies (especially in their harsher monolithic phases) add further features of their own.

North America, as a prototype of a modern industrial society, seems to present the following combination of images and arrangements. First, individuals, in their comparable rights to the opportunity for accomplishment and the pursuit of private fun, matter more than collective obligations. But individuals are declared to be comparable. Second, the criteria for appraising individuals imply a kind of accomplishment that can thrive neither on native talent alone nor on self-cultivated personal qualities. Instead, they demand training and schooling—the training and schooling appropriate to specialized achievement and free from fanatical commitments to ideologies. Youth is to be educated by hortatory and certified samples of the adult population who teach in a professional spirit. Youth is not to be systematically capitivated by charismatic teachers who could not be easily replaced or who would reduce the relatively easy transfer of students from one classroom or school district to another. Youth is to be prepared for a future different from the present, by adults who expect to be displaced and overtaken by youth.

The adult world, moreover, sees in youth a special embodiment of its own wider assumptions: that life is open, that there is space for change and betterment, and that the future will improve on the past. The adult world is antitragic, antifatalistic, and in that sense, it is young. Yet this fact places a special burden on youth, for youth is even younger and seeks both guidance and autonomy. Besides, the adult world facing youth is neither a coherently educative (if differentiated) reservoir of models, guides, and supports, nor is it free to stand detached, allowing youth to identify with any one sector of the adult world against another. Instead, youth tends to stand equidistant from home and school as it seeks emancipation for its early dependence; and thus youth is drawn into private reciprocities with its peers that are collective or even more intimate.

The burdens that its early dependence impose on youth are after all demanding, and they have been made so by at least two phenomena: the special attention we give to children, and the rather high degree of self-sufficiency we exact of adults. Some incongruous facts result. As part of the general movement toward greater equality, which in turn is influenced by the corrosion of views that justify inequality, we have become increasingly concerned with doing well by our children. We are prepared to argue that they deserve only the best, and our concern, which has displaced certain gross forms of cruelty, is accompanied by certain modern forms of parental anxiety and uncertainty. Still, we seek to be generous and gentle with children. We have probably forgotten that the original meaning of "raising" children came from *tollere*, that is, the father's lifting the child from the ground, as a sign that he wished it to live, not die by exposure or otherwise. Children, therefore, have become deeply engaged with their parents, while later, as parents themselves, they must live on their own.

Thus, in the West, we have to some extent shifted the children's burdens onto youth. Youth now symbolizes the contradictions generated by our values and practices. Further, they live these contradictions; they remind us, their elders or progenitors, of our own contradictions, and they demand, often inarticulately, that we help rectify the delayed costs of our early but unsupported wish to understand our children. It is as though childhood were the beneficiary of the more cooperative and responsive elements of our norms, while the more austere, even ruthless dispositions are held in abeyance. Yet these are present, and what was begun in childhood cannot for many reasons be sustained. Ultimately, both generations come to feel that the one has let the other down. In this respect, adolescent youth, with its demand for personal experience, for charismatic leaders that elicit devotion and intermittently fanatical phases of involvement or withdrawal, undergoes the experience of betrayal. At the same time, its elders are reminded that the contradictions of the beliefs without which they cannot act have been temporarily exposed—precisely when they see youth leaving childhood to join the ranks of the adults. After twenty, as we know, absolute differences in age can become progressively larger, while at the same time they often make a diminishing difference.

In addition, the very requirements of the industrial and hence technically dependent society, almost despite itself, have made "youth culture" a literal fact. Change and obsolescence are exceedingly evident. "Continuing" education, retraining, and refresher courses symbolize the acknowledged fact that for many of us the formal period of learning cannot easily come to a full stop, to be followed by a long stretch of "adult" performance. In order to be continuous, learning, now associated with youth, will have to become compatible with adulthood, and, in a society committed to sustained technical change, it must be continuous. Learning, then, will not mean a distinction between youths and adults. Thus, in the West, some of the discontinuities between youths and adults will have been dissolved.

Meanwhile, especially in America, youth's concern with personal involvement may take the form of a search for "security," though in the main this seems much too simple a formula; and this concern is a further reminder that, for all the adults' emphasis on the qualities of youth, the major commitments of North American culture, until recently, at least, have moved us away from personal intensities in order to keep us sober and free enough for impersonal accomplishments. (Periodic intoxications, whether alcoholic or other, only confirm this pattern.) In that respect, then, the very culture that fears old age and is troubled about aging and that values youth often also lacks the inner space necessary for the cultivation of personal experiences and licit adventures which, pursued in vigor, testify that one is indeed (still) young. It is no accident that ultimately, in such circumstances, youth is considered a temporary fact. Dreams have to do with eternal youth; but in life, youth looks like a finite sum. In retrospect, it will have been well or poorly spent. Such economic language links youth on the plane of values to the subsequent stage of adulthood. In actual experience, however, these two stages contain important discontinuities.

In totalitarian societies, in which the important balance between the public and private domains does not exist, youth can be harnessed into movements. This is always possible wherever the ties of kinship, or of the nuclear family, are relatively weak, while the demands of the encompassing political community are strong and overriding. In these circumstances, the isolation of youth appears to be less than in our own case. Semicharismatic leaders give this youth its bonds with its peers. Quite apart from the loss of personal freedom, however, the existing order (whether originally revolutionary or not) suggests to youth that what the believing elders represent to them is in fact to be believed. Meanwhile, the phenomena of bureaucracy, science, and technology are transcending ideology and making past history in certain respects irrelevant. These phenomena confront youth with certain uniform, ever more widely distributed tasks—tasks of training themselves so as to adjust the sooner to the impersonal social arrangements that seem to mark the end of childhood. It may well be, then, that through an earlier adulthood we are already dissolving some of the pains of youth, and that in turn the adults will become the outsiders.

In summary, I should like to put forward these suggestions, without attempting to prove them. In America (more in the United States than in Canada), we have been very conscious, and proud, of being a young society. Our image of youth has been especially concerned with the qualities of vigor, enterprise, and a certain purity of motive. Characteristically, this image has excluded a commensurate emphasis on the intensities, the absolutisms, and the shifting moods of youth. When these are actualities, exhibited by actual youth, they often constitute sources of strain and displeasure among the very adults who value "youngness" (if not young people), and who are eager to hide their own age.

In a complex way, we may still believe that "youth should not be spoiled by old age," but we certainly do not believe that "the best gift of the ageless gods is to snatch it away." Instead, we hope that we shall persist in achieving, but that in doing so our age will not show, and that our youths will do likewise. Yet our youths have come to receive the harsher side of our values, while our children receive the gentler. Meanwhile, the process of industrialization is generating certain common conditions of life that are dissolving the previous contradictions in moral and ideological commitments, while at the same time they are reducing the opportunity for legitimate intensity. In the West, this process tends to leave youth in an isolated position.

Under totalitarian conditions, especially in the early phases, when the regime is still young, youth can be enlisted collectively—and is of course forced to enlist. This isolation of youth is both a cause and a consequence of our wish to combine freedom from extended familial and other inherited loyalties with freedom from a persistent and dominating concern with political obligations. The isolation of youth proceeds, however, in the presence of extended opportunities for adults to participate in civil matters. These facts are in turn balanced by a pervasive sense that the individual is powerless in the face of a putative widening of the bureaucratic patterns that veil the seats of power. Youth, in reaction, often becomes passive and withdrawn— one form of rebellion, which is also a demand for autonomy. On the other hand, the cultivation of private experience is consonant with certain major values of the society within which this rebellion ambiguously takes place. Thus adults, in their stress on youth, are often disappointed in youths themselves, while the latter find their elders frightened by the conflicts youths present and all too often unhelpful in resolving these conflicts. In both youths and adults, therefore, there arises a feeling of having been let down. The ensuing mild despair is then converted in part into a subsequent hopefulness for children. Thus the spiral continues.

REFERENCES

1. Max Wegner, *Greek Masterworks of Art,* translated by Charlotte La Rue (New York: George Braziller, 1961), p. 24.

2. *Ibid.,* p. 25.

3. Wolfgang Borchert, *Das Gesamtwerk,* with a biographical note by Bernhard Meyer-Marwitz (Hamburg: Rowohlt Verlag, 1949), pp. 59, 60-61.

4. Michael A. Wallach and Leonard R. Green, "On Age and the Subjective Speed of Time," *Journal of Gerontology,* 1961, *16*: 71-74.

5. Alice Meynell, *Children of the Old Masters, Italian School.* London: Duckworth & Company, 1903. C. Haldane Macfall, *Beautiful Children Immortalized by the Masters.* London: T. C. and E. C. Jack; New York: Dodd, Mead, 1909. Asher Tropp, *The School Teachers* (London: William Heineman, 1957), pp. 57, 184.

6. Jakob Burckhardt, *Die Kultur der Renaissance in Italien,* introduction by Wilhelm von Bode (Berlin: Th. Knaur. Nachf., 1928), pp. 208-209.

7. John Christmas Møller and Katherine Watson, *Education in Democracy.* London: Faber and Faber, 1944.

8. Friedrich Paulsen, *Geschichte des Gelehrten Unterrichts* (Leipzig: Veit & Company, 1919), vol. 1, pp. 640-649.

9. Michael Robert, *Gustavus Adolphus: A History of Sweden, 1611-1632* (London-New York-Toronto: Longmans, Green, 1953), vol. 1, ch. 8, pp. 428-441.

10. Marion Levy in *The Family Revolution in Modern China* (issued in cooperation with the Institute of Pacific Relations; Cambridge: Harvard University Press, 1959) provides a detailed discussion of the "absolute age groups" (pp. 66-133) through which an individual passes, as well as of "relative age considerations" (pp. 134-147). We know from our own society how these considerations can produce incongruities when an uncle, for instance, is much younger than his nephew. One's age, in other words, puts one into one irrevocable position after another, but one's age is also relative to the age of others.

11. Kaspar D. Naegele, "Clergymen, Teachers, and Psychiatrists," *The Canadian Journal of Economics and Political Science*, 1956, 22: 46-62.

BRUNO BETTELHEIM

The Problem of Generations

WHAT STRIKES the psychologist forcefully when he surveys the available literature on adolescence and youth is that, if the amount of discussion were indicative, then all or nearly all problems of youth would appear to be those of the adolescent male. True, the more serious authors nod in the direction of female adolescence and recognize that it creates problems, too. But having done so, they turn so exclusively to the problems of the male adolescent that the net impression remains: female adolescence, if it exists at all, does not create problems equally worthy of the sociologist's or the psychologist's interest.

But whether we view adolescent development from a sociological or a psychological viewpoint, the problems confronting boys and girls should be parallel. The reassertion of sexual desires on reaching physical maturity is typical of both sexes, as are the psychological problems of repressing or satisfying these drives; of postponing the consummation of some and of sublimating the rest. So also, in modern times, are the social and psychological problems of achieving self-identity on a more mature basis, and of finding one's place in society.

Is it really so much easier for the adolescent girl to find her self-identity as a woman and her place in society, than for the boy to gain his as a man? True, Erikson describes cases of negative identity in women, but most of his writings on adolescence concern males, and other recent students of the problem such as Friedenberg and Goodman deal almost wholly with male youth.[1] Though Freud's original work was based on the study of hysteria in females, and though he devoted much thought to their difficulties in achieving sexual maturity, a problem typical of middle-class youth, his later writings centered mainly on the development of the male. In another context I have discussed how, in regard to puberty rites, nearly all psychoanalytic interest centers on boys, neglecting the far-reaching meaning the rites have for both boys and girls in achieving sexual maturity and adult status.[2]

Since I, too, have had to rely on available sources, and since they are so much richer in content and more abundant than any one man's observations can be, my discussion too will be weighted toward male youth; but at least I wish to acknowledge this deficiency in my remarks, and try to rectify it in part. I venture to say that those who conceived of this issue of *Dædalus* were caught in much the same predicament, since a careful reading of the suggestions they kindly offered me permits a consideration of the feminine only by stretching the points they detailed. Yet there must be a reason why the male adolescent and his problems dominate public attention and that of the scholarly expert.

Perhaps my particular topic allows for a first approximation as to why this is so: I was asked to discuss the problem of the generations from the psychologist's point of view. Unfortunately I shall have to transgress heavily into the field of sociology, since in my opinion the problem of generations is at best a psychosocial one, and can never be dealt with on a purely psychological basis.

It may be that the problem of generations is what gives us adults so much trouble, and not the problems of adolescence or youth; and this is why, when we concern ourselves with the problem of youth in our society, it is that of male youth. If delinquency worries us, it is chiefly male delinquency, though to their families and themselves, female delinquents are at least as great a problem.

For the same reason we are concerned with Johnny's not learning to read, or not getting enough science and math, as if reading problems were foreign to Jane, or as if she were automatically good at math and science. And if not, it is thought to matter little, since the number of female contributors to the sciences is relatively negligible. But such an evaluation of women's potential is both short-sighted and wasteful. Despite all the obstacles to women's higher education in the physical sciences, it was a young woman who did pioneering work in radioactivity, a development in physics which eventually led to the present clamor for more science education in our schools.

Since it is adults who conduct the studies and write the articles, and since these adults are predominantly male, they write most of where the shoe pinches them—that is, in regard to the problem of generations. They are neglectful of where the shoe hurts the young —that is, with the problem of sexual maturity and with finding one's place in society. True, each of these is well recognized as problematic, but especially so where they coincide with what is bothersome to adults, the relation of the generations to each other. They are neglected where they trouble adults less, as in the female.

I hasten to add that this respite in regard to the girls' part in the problem of generations is fast disappearing. I submit that already it creates more emotional hardship than that of the males, which at least is officially recognized. Why, then, is the problem of generations so much more acute in the male than in the female?

To put it crassly: the self-identity, and even more the self-realization of the young man, implies to a large degree his replacing the preceding generation. In order to come into his own, the old man (or whoever stands in his place) must move over; or, in the folklore of my native Austria, he must move into the old people's quarters (*Ausgedinge*). This happens as soon as the son is ready to take over the farm, and with it, all other prerogatives including the main building: the farmhouse.

Such ascendancy of youth over old age proceeds smoothly, at the right age and in the correct form, if it tallies with the survival needs of the entire family and with the facts of biology, in short, if it takes place in concordance with nature and nurture. The well-being of the farm family depended on a vigorous male being in charge at least of the farming, if not also of the family. For centuries most men lived on farms tilled by small landholders or serf-tenants, or else made a living as small artisans or shopkeepers. In those times economic success and often mere survival depended on the physical strength and skill of the head of the family. So it seemed "natural" that as the father's vigor declined, a son just reaching the prime of his strength and mental abilities should take over.

When this was so, and since the father's life was meaningful, that of the son who at first helped him and then followed in his footsteps was automatically meaningful, too. An old Chinese proverb summed things up: "He who has sons cannot long remain poor; he who has none cannot remain rich." This was something both fathers and sons understood. The son growing up was secure in everyone's knowledge that he added substantially to the economic well-being of the family. Seeing also that his contributions increased in importance as he approached maturity and the peak of his physical strength, just as his father was declining in vigor, he had no need to worry about his work achievement (as would now come through more indirect employment on the labor market) and whether it supported his claim to have reached manhood.

Such an easy succession of the generations, even in times past, was mainly an "ideal" solution to the changing of the guards. But actuality often approached this ideal; and if not, the biological realities made it seem like the given order of things, since production depended so largely on the male's prowess.

True, even when the level of technology was so low that physical strength counted most, one's experience and skill, the know-how of work and of life, had to be added to strength to succeed; but all one's experience and knowledge were of little avail without physical strength, because it alone powered the economic process. There was no point, therefore, in a selfish holding on to knowledge or even to property rights, because they were only theoretical if they could bring no return, once physical power had failed.

While there were always some old men who held on beyond reason, along with other conflicts of interest and generations, these were clearly men smitten with blindness or carried away by unreasonable emotions. An intelligent self-interest in both old and young still required a transition of power and privilege at the point when physical decline set in among the aging and manhood was gained by the young. An old man might stand in the way of his sons, but his fate was then ordained. Thus O'Neill's Ephraim Cabot is unwilling to recognize his son's contribution; but when the son walks off, not with the farm but with his father's young wife, the father ends up in possession of a farm that he alone cannot tend to.

It is when physical strength is no longer essential for survival or economic success that the biological process of aging as well as of maturing no longer, of necessity, conditions the taking over of the dominant position by youth. What once formed the "ideal" solution, because conditioned by the nature of man, suddenly turns into an arbitrary "ideal" without any necessary or natural basis. Such an ideal soon becomes hard to put into practice, and eventually even ceases to be an ideal.[3]

Even before modern times, wherever physical strength was not essential to survival because the life-assuring labor was performed by others, the problem of generations was acute. This was true, for example, of former ruling classes, and later on of the upper middle classes. It was also true in the Greece of Alcibiades and in Catiline's

Rome. Whenever there was no natural order to the ascendancy of the generations, problems arose between them similar to those that are now typical for all in a machine age, when almost nobody's survival depends on physical strength.

If the young man's coming into his place as head of the living unit is not thus assured by the natural order of things, if he cannot be sure that the dominant position will be his at a foreseeable and not too distant moment, then he cannot wait for it in good grace. Then he must fight for it, for both his rights and his obligations, and the sooner the better, because only both in their combination make for the realm of the mature man.

If such a transition does not occur smoothly, is not accepted as natural and inevitable by both partners, then the older generation is likely to view the younger with suspicion, and justifiably so, because youth taking over is no longer necessary and natural. Why should the older generation voluntarily abdicate if it has nothing to gain by it and loses nothing by holding on?

Many if not most adults have an emotional need for children and enjoy bringing them up. It is such a truism that children for their part need their parents for physical and emotional survival that I mention it only to round out the picture. Once childhood is past, however, the picture changes. At certain times in history the older generation had an emotional need to see its way of life continued by the coming generations. This was particularly true when the parent generation had begun a work it could not complete and which it felt would remain pointless if uncompleted, be it the clearing of the land or the raising of a cathedral. Yet the more we came to doubt that things would continue in the old ways, that we were toiling for eternity, the less emotional need was felt by the older generation for the next one to continue what it so auspiciously had begun.

With the advent of modern technology and mass society, only very few have so intimate a feeling for their life's work, such a personal investment in it, that they need to see it continued by others. Short of such a desire, the older generation has little psychological need for youth. If youth tends to move away and build a life very different from that of the old folk, whom they only sporadically remember or visit on special occasions, then even the hope for emotional comfort from youth becomes unrealistic. While youth may still have some emotional and economic need for parents, most parents have little emotional need, and very few an economic one, for a youth striving to be free of its elders. It is because parents still have an emotional need for children, but not for an independent youth, that they often show strenuous resistance when youth fights for its independence. It is also what makes them so critical of certain exaggerations or passing effects of youth's battle for self-realization.

Such resentment and ambivalence about youth's striking out on its own matters little if youth can readily remove itself from the impact. The development of youngsters who went West or ran away to sea was not hampered by adult criticism of their ventures, though their development to full emotional maturity may have known other vagaries. But if youth stays at home or close to home and still fights for its independence from those it depends on, both sides show an emotional deficit.

The resultant scarring of personality in a delinquent youth, for example, is recognized by society as serious. The misanthropic nagging and dissatisfaction of his elders are less well recognized as the price they pay for the conflict. Hamlet thus has his counterpart in King Lear, who, unwilling to make room for youth, tries instead to put reins on the younger generation and to saddle it with a burden of gratitude. It is poetic justice that Cordelia, willing to serve age by foregoing the right of youth to a life of its own, suffers destruction, too.

In the psychoanalytic literature certain aspects of the problem of youth are traced back to a revival and a more violent acting out of the oedipal situation in adolescence. And it is true; something akin to the oedipal situation may be found among the children of most known societies. Specific variations will depend on who is head of the family and to what degree; the character of the persons who minister to the child; how large or small the family is; and how intense or weak is the emotional attachment of specific family members to one another and to the child. But the same cannot be said for the repetition of oedipal conflicts in adolescence.

The girl who marries at fifteen or sixteen and soon thereafter has children of her own is not likely to be beset by a repetition of her oedipal longings for her father or by any fierce competition with her mother for his emotions. As suggested earlier, the repetition of the oedipal conflict is not an issue of nature but depends very much on the structure of family and society.

Only a youth who is kept (or keeps himself) economically and emotionally dependent on the older generation will experience the repetition of the conflict that psychoanalysts observe so frequently nowadays among middle-class adolescents. To cite Hamlet again as the most familiar instance of a revived oedipal conflict in youth—if Hamlet, like Fortinbras, had fought for a kingdom across the sea instead of wishing to inherit his father's place, no tragedy would have taken place; but because he wished to take over from a generation unwilling to yield, rather than to find and win a world of his own, the old oedipal feelings were reactivated and led to the tragedy that destroyed them all. If Hamlet had known for sure what he wished to achieve on his own in life, he could not have been pressed into becoming the avenger of his father. For Hamlet's father, like Lear, put a private burden on his child's too weak shoulders.

Here, then, is another aspect of the conflict of generations: the parent who sees his child's main task in life as the duty to execute his will or to justify his existence (which is different from a parent's devotion to an unfinished labor which the child, on his own, later wishes to bring to fruition). The son who does not revolt when he is expected to devote his life only or mainly to achieving what the parent could not, usually perishes as Hamlet did.

In the present-day world with its tamer middle-class society, we find that the conflicts between a youth either afraid of or prevented from coming into its own and an older generation unwilling or unable to give way, are no less tragic though played out in more muted tones. It follows that whenever society is so organized that youth remains dependent on the older generation, because of the duration of the educational process or for other reasons, and this older generation is not ready to step aside economically, politically or emotionally, a psychological impasse is created which may then be aggravated by unresolved oedipal conflicts.

Here it might be well to remind ourselves that no oedipal situation would exist if the parents were not deeply involved with their child. The revival of the oedipal conflict in adolescence is often due to a parents' wish or need to remain as important to his child in adolescence as he was during infancy.

I venture to guess that many more (particularly middle-class) youths come to grief nowadays because of their parents' insistence that the former justify them as parents than because of any revived oedipal desire for their mothers or fathers. (This again is different from youth's independent wish to prove its own worth and not the worth of a parent.) One form of such an insistence is the overt and covert pressure on youth to provide the parents with what was lacking in their own lives. The mother's pressure on her daughter to live out vicariously her unfulfilled daydreams of popularity or to make a notable marriage can effectively block the girl's efforts to find self-realization in ways that are genuine to her. Often it is both parents who expect their sons to excell in athletics and who take for granted their right to assume that their child will do better than they did.

Nowadays it is usually both middle-class parents who put pressure on youth of both sexes to enter the prestige colleges; this pressure is reinforced by a parallel one from the schools and the general public. I have found those parents most insistent and most unreasonable in such demands who never went to college themselves or never graduated.[4] Many a college youngster needs to ward off this undue attempt to run his life as his parents or teachers want, and longs to carve his own way. He decides that the only way to manage this is to drink, do poorly in college, or flunk out. That is not his original desire; he acts out of a necessity to prove himself master of his own fate.

All this is only part of an attitude that expects American children to do better than their parents, and often, seen objectively, the task is even quite feasible; but to children and adolescents the demand seems emotionally impossible, because it comes at a time when their opinion of their parents' achievements is unrealistically high. Contrary to all psychoanalytic writings that teach clearly how the child and adolescent is overawed by his parents' power and wisdom, both society and his parents continue to expect the emotionally impossible of youth. Off-hand, I can recall no single statement in which consideration of what is expected of high school and college youth is directly linked to the achievement of the youth's parents. While we are more than ready to praise the self-made man, we are reluctant to apply the correlate of such praise: to recognize how difficult it is to outdo one's parents.

True, many youngsters end up doing better than their parents, either socially, economically, or intellectually; but I wonder how the score would show up on a balance sheet that also took account of emotional well-being. A vast number have risen in this way, but then they have hardly been able to manage life, even with the help of a psychoanalyst. If we consider, in evaluating those who do better in life than their parents, not only the externals but also the inner life, perhaps the picture of success that emerges would give us food for second thoughts.

To put the burden of surpassing one's parents on the relations between parent and child leads of necessity to unresolvable conflicts. If youth succeeds, it emasculates the parent. As a result, youth cannot feel successful—partly out of guilt, and partly because he cannot be sure if it was he or his parent who wanted him to succeed. That is, he cannot be sure who it is that truly structures his life.

As for the older generation, the conflict shows many faces. It may take the form of contempt if youth does not fight back (they are weak) or of hostile anxiety if it does (they are delinquent). And if youth has serious doubts about whether it will ever succeed, it must still either rebel or submit in cowardly fashion, or else find some devious (neurotic) way to sidetrack the issue. (To be neither son nor man avoids the fight altogether and hurts the father most, as when the son is a "beat.") Hence the eternal historic and dramatic predicament of the crown prince: if he submits to his father's superannuated clinging to office, he will be a weakling when he finally inherits the throne. If he rejects a role of empty waiting, he must head the revolt against his father.

But what about the girl? Must her mother abdicate for her to come into her own? Not as long as her psychosocial identity resides in childbearing and homemaking. The older generation may stand in her husband's way of realizing his independence, but as long as she accepts that her independence as a social being rests on his, the older generation may stand in her husband's way but it does not stand in hers. Her mother does not need to move over for her to be herself. On the contrary, the mother's having reached an age when childbearing is no longer possible or becoming makes it obvious to one and all who is now in ascendance, and no fight between the generations is needed to settle the issue. (Some modern mothers who cannot accept their gray hair and fading looks, and with it their sexual decline, create a problem for their daughters similar to that of the boy's.)

But what if the girl's psychosocial identity ceases to reside in childbearing and homemaking, or exclusively so? Until the industrial era, as Veblen saw, woman's social identity was "essentially and normally a vicarious one," an expression of the man's life at second remove. And so long as she remained (of necessity) a drudge, she accepted this ancillary role and was largely at peace with her lot. But the less this became true, the more her problems of identity and self-realization were compounded. By now, the female adolescent struggles not only with having to decide whether her place is in the home or in society at large, or in both, but to what degree and with what justification. Thus the problems of youth have become nearly the same for both sexes; the sexual difference counts for less, because the conflicts of growing up are so much more psychosocial than sexual.

This, I believe, is one of the reasons why psychoanalysis is so often ineffective in adolescence—not because the sexual pressures are so great, and they are great, but because psychoanalysis, which is so well able to help with problems of sex and repression and personal self-realization, does not help with the problem of social self-realization. Or, to put it differently, pitting a helpful authority (the analyst) against repressive authority figures still leaves the adolescent under the sway of some adult authority which he needs to replace with his own. Or, to put it yet differently: psychoanalysis is devised for and effective in helping persons with their intrapersonal difficulties; hence it tends to approach all problems as such. But the problem of the generations is an interpersonal difficulty. Therefore, to deal with it as if it were intrapersonal only complicates matters instead of simplifying them, and makes resolving them less likely.

Of course, this also is true only for Western middle class society. How the problem of generations can differ in different cultures may be illustrated by a controversy between American and Japanese psychoanalysts: in Japan the psychoanalyst's task was seen to consist in helping the young individual to give up his search for self-identity; his self-realization was to be sought not in individuation but in accepting his place within the family in the traditional subservient position of the son toward his father. Thus a Japanese patient "as he approached the successful conclusion of his treatment said, 'During my vacation my mother told me on one occasion that I was now pleasing my father better again.' The psychoanalyst, in reviewing the changes in the patient's personality, says, 'His psychic state is now as harmonious a one as can ever be reached by human beings' i.e., in accordance with the national mores and aspirations of Japan."[5]

Most serious writers on the problem of youth have recognized that youth's present difficulties in Western society are closely related to changed social and economic conditions and to the ensuing difficulty for youth in finding self-realization in work. As Goodman observes: "It's hard to grow up when there isn't enough ⌐ and he continues, "To produce necessary food and shelter is ma⌐ work. During most of economic history most men have done this drudging work, secure that it was justified and worthy of a man to do it, though often feeling that the social conditions under which they did it were not worthy of a man, thinking, 'It's better to die than to live so hard'—but they worked on. . . . Security is always first; but in normal conditions a large part of security comes from knowing your contribution is useful, and the rest from knowing it's uniquely yours: they need you."[6]

Just as in this country an earlier generation needed youth because the economic security of the family depended on its contribution, so in Russia today youth is needed because only it can carry on the task of creating the new and better society; and in Africa because only it can move society from tribal confusion toward modern democracy. If the generations thus need each other, they can live together successfully, and the problem of their succession, though not negligible, can be mastered successfully. Under such conditions youth and age need each other not only for their economic but even more for their moral survival. This makes youth secure—if not in its position, at least in its self-respect. But how does the parent in modern society need the next generation? Certainly not for economic reasons any more, and what little expectation a parent may have had that his children would support him in old age becomes superfluous with greater social security. More crucially, the status-quo mood of the older generation suggests no need for youth to create a much different or radically better world.

In many respects youth has suddenly turned from being the older generation's greatest economic asset into its greatest economic liability. Witness the expense of rearing and educating youth for some twenty or more years, with no economic return to be expected. Youth still poses emotional problems to the preceding generation, as of old. But in past generations these emotional problems were, so to speak, incidental or subservient to economic necessity. What at best was once the frosting on the cake must now serve as both solid food and trimmings—and this will never work.

Thus the economic roles, obligations, and rewards are no longer clearly defined between the generations, if not turned upside down. Therefore, another aspect of the relation between the generations looms ever larger, in a balance sheet of interaction that is no longer economic but largely emotional. Modern man, insecure because he no longer feels needed for his work contribution or for self-preservation (the automatic machines do things so much better and faster), is also insecure as a parent. He wonders how well he has discharged that other great function of man, the continuation of his species.

At this point modern youth becomes the dreaded avenging angel of his parents, since he holds the power to prove his parents' success or failure as parents; and this counts so much more now, since his parents' economic success is no longer so important in a society of abundance. Youth itself, feeling insecure because of its marginal position in a society that no longer depends on it for economic survival, is tempted to use the one power this reversal between the generations has conferred on it: to be accuser and judge of the parents' success or failure as parents.

How new is all this? It is very hard to compare one age with another. But the Alcibiades or Catiline of antiquity would not have had their followings if the problem of youth having to test itself against an older generation had not existed in those times; nor do Plato's indictments of what he saw as obstreperous youth sound very different from those leveled at our young people today. I may be the victim of those distortions of perspective that make things distant seem far smaller than those looming in the foreground.

Whether this is error in judgment or not, the fact remains that the present problems of Western youth in finding self-definition, and with it security, seem more complex than those of other generations. I say Western youth because, while Russia appears to have its equivalent of the Teddy boys and while Israel does not seem altogether happy with all aspects of Kibbutz-reared youth, the problems there seem different not only in quantity but also in quality. The main difference lies not so much in the particular tasks society sets for its younger generation but in how clearly the latter realize that only they, the generation of the future, can achieve these tasks.

This difference is critical, for, contrary to some people's opinion, youth does not create its own cause for which it is ready to fight. All it can do is to embrace causes developed by mature men. But youth can only do this successfully if the older men are satisfied with providing the ideals and do not also wish to lead the active battle for reaching them. Or, to put it differently, a youth expected to fight for his personal place in a society of well-defined direction is not lost but on his way. A youth expected to create a new but not yet delineated society finds himself a rebel without a cause. Only when each group has its own important tasks, when one without the other cannot succeed, when age provides the direction but youth the leadership and the fighting manpower, is it clearly understood that whether the battle is won or lost depends on youth's fulfilling its all-important share of the total struggle.

As to who is to provide meaningful work for youth, I believe the answer is that nobody can do that for another. This is why I believe that the well-meant discussions and advice as to what industry should do to make factory work more meaningful has the problem all wrong. Nobody can make life or work more meaningful for others. Nor do most tasks have an absolute significance, not even the growing of food.

On our own plains, for example, stand acres of corn cribs filled to overflowing, while youth continues to leave rural America in droves. Yet its first response to the contemplated Peace Corps for underdeveloped areas was electric. This was the more striking since American farms are now largely mechanized (it is clean work), while the very goals of the Peace Corps include the mechanization of agriculture wherever it is still largely manual (which is "dirty" work). So it is not the work task (growing food) which attracts or repels youth, but the clear evidence that "they need you."[7]

Yet when it became apparent that to join the Peace Corps candidates had to be screened, take examinations, and then be assigned tasks, enthusiasm faltered. It was not initially aroused by the chance to enter one more rat race of competitive examinations nor by the prospect of being sent where the managers of the Corps wanted them to go. They did not want to be emissaries, even of their country. They had jumped at the chance to go where they were needed and to prove how well they could do. They hoped to develop themselves while helping others to develop their country. When it became clear that they were expected to represent something else (in this case American goodwill abroad), they lost interest and, of the many who initially applied, few presented themselves for the scheduled examinations. A chance to act on their own fervor caused a stir; but faced with one more competition, they might as well continue the college rat race for suburbia.

True, a society of plenty can tempt people to waste their time by filling it with empty entertainment or meaningless comfort, or to strive for the wherewithal to do so. But no one need fall for this temptation nor can even the best TV programs make life more meaningful for viewers. At best they can provide the raw

76

material which the individual can then forge into a meaningful life.

Just as freedom and democracy cannot be handed down but have to be fought for, just as knowledge cannot be poured into the heads of our students but only situations created that induce them to seek it, so too industry cannot make work more meaningful for the worker. Only he can first find out what kind of work may be meaningful to him, and then go out and seek it, or at least a reasonable compromise between what is personally meaningful work and what jobs are available to him. I think those who complain that work is not more meaningful are in the wrong, too. What is wrong is that more people do not strive to find meaning in their lives; if they did, they would radically alter our economy and with it our working conditions.

Thus the older generation never has provided meaningful work or life for youth. All they have striven for was a deeper meaning in their own lives, and when they did that, segments of youth could at least follow their lead. It was the mature Marx, not the adolescent, who created Marxism, which then provided the basis for meaningful effort by a whole generation of youth. It was not the youthful but the mature Roosevelt who stimulated another generation of youth to find meaning in life through social improvement and through efforts to reorganize society.

Surely, we too give lip-service to the conviction that man's best hope is the next generation, but this hope does not seem very strong or attractive if we of the older generation do not pursue it with equal vigor. Neither our conviction that the West is declining nor our fear that atomic destruction will wipe man from the earth, realistic as each may well be, offers much hope for assertive self-realization, now or ever. If I cannot feel myself full of vitality because of my hopes for a life in the future, if the world I am about to create will not be better than that of my fathers, better not to live in this world, better to retire from it or feel alive in the moment, no matter what price I must pay in the future. So reasons the criminal delinquent who seeks a moment of heightened self-assertion in the anxious excitement of the criminal act, or through the kick he finds in his drugs.

As one delinquent youngster complained, "You can't live, if there's nothing to push against." What he meant is that you cannot test your own worth, your own strength and vitality, the very things you feel most dubious about as an adolescent, when all you can push against is a vacuum, or an adult society more than ready to give way, to act more youthful than even befits youth. Without something definite to push against, youth feels lost. Many causes are embraced by youth, not for the cause itself, but because in fighting for it, its strength can be tested against something. Hence youth favors causes that run against the established order, even an ultra-conservative cause, because nothing is quite so safe a testing ground as the well-established order.

In Germany today, delinquent youth is often spoken of as the half strong (*die Halbstarken*). They are half strong because for them the older generation and its values mean the Hitler generation. But the ideas and values of those elders proved deficient in all important respects, while the generation of the fathers was cut down by heavy German losses during the war. Hence youth could not test itself against them and remained only half strong.

In the United States a very dissimilar but parallel process took place. The depression led to serious doubts about the values and ideas of the older generations as to the merits of a free enterprise society. The response of the older generation was frequently an inner abdication of the truly parental role. Since that generation felt it had embraced the false goddess of material success, they relinquished being mentors of the next generation and tried instead to be their pals, if not also their peers. They did not, however, give up wanting their children to give meaning to their own now emptier lives. While German fathers were either absent, or died in the war, or were morally destroyed by becoming Hitler's servants, many middle-class American parents simply abdicated their parental function but still wanted their parenthood proved successful by their children's achievements in life.

In terms of generations, then, the question that haunts every young person is: am I as good, as much a man as my father? as much a woman as my mother? This is something to measure up against, to find out if one has it in him to push things a bit further than the parents were able to do. But if the question has to be: am I as much a man as my father should have been, as my mother wants me to be? Or in reverse, and hence even more void of direction, the anxious question: am I the girl my father does not want me to be? Then the person is lost, without guideposts in his struggle to find out what kind of person he is and what kind of a person he wishes to be, as compared with the generation of his parents.

Where is youth to go? How is it to shape itself and its relation to the older generation, the image it must either want to emulate or to supersede with a better one? If I am not mistaken, it is in Jack

Kerouac's *On the Road* that two beat characters have the following conversation: "We got to go and never stop till we get there," says the first. The other wonders, "Where are we going, man?" and the answer is, "I don't know, but we gotta go."

These two young men are not in flight from society. They seek a goal—that much is clear. Otherwise, the first would not say they must go until they get there, nor would the second ask where they are going. If they were merely in flight, the first might have said they must go and never stop, and the question of destination would not have come up. More than that, they are in a great hurry to get started toward their goal; but this goal is elusive, and so they are people lost in their search, so lost that they no longer know which direction to take. Worse, they doubt that there is any direction. Therefore, their search for only an unknown goal becomes empty roaming. As long as they are on their way, they feel alive. If they stop, they fear to die. Therefore, any and all kinds of spurious activities will do, to keep from recognizing how lost they are.

Why is this goal eluding modern young man in search of himself? If manhood, if the good life in the good community, is the goal of adolescence, then the goal is clear, and with it the direction and the path. But what if existing manhood is viewed as empty, static, obsolescent? Then becoming a man is death, and manhood marks the death of adolescence, not its fulfillment. The bouyancy of youth is fed by the conviction of a full life to come, one in which all great things are theoretically attainable. But one cannot believe in the good life to come when the goal is suburbia. One cannot realize one's values by climbing the ladder of the business community, nor prove one's manhood on the greens of the country club; neither can one settle into security in an insecure world.

If there is no certainty of fulfillment, then it is better not to give up the promise of youth with its uncertainty, its lack of definite commitment. Youth at least offers a chance to escape the premature death of rigidity or the anxious confusion of a life that is disgraceful when it is without direction. Neither rigidity nor a confused running in many directions at once (and running after status or money are only the worst among nondirections) is an attractive goal for the young man trying to emerge from his state of uncommittedness into one of inner stability. Better to be committed to such uncommittedness than to commit oneself to spending the rest of one's life as a hollow man.

One's fathers (at least, the best of them) did a good job of showing the young how hollow a life they had built. Let this be a warning not to join them in the waste lands. Better not to enter this land of walking shadows, of immaturity posing as maturity. Better to assert defensively one's uncommitted immaturity, one's remaining poised at the threshold of a life one does not wish to enter. It is the romantic position, but alas, the position of a generation which has little belief in the romance of a better world, a generation whose dreams are not to be striven for but subjected to analysis.

I have said that the present difficulties of youth are related to changed economic and social conditions, and to how much harder it has grown for youth to find its fulfillment in work. In this respect the fate of the girl can be even harder than the boy's. It is impressed on her from an early age that her main fulfillment will come with marriage and children, but her education has nevertheless been the same as that of boys, who are expected to realize themselves mainly through work and achievement in society. To make matters worse, the years in college or even graduate school have further prepared the female elite to seek self-realization in work, while society at large continues to stress that they must find it in motherhood.

Only very occasionally, for boys, is fatherhood added like an afterthought as part of their self-image as mature men. And nowhere, to the best of my knowledge, or only most incidentally, is the complementary image of being a husband even dimly outlined. Yet it should be obvious that women will not find fulfillment in being wives if their partners do not see being a husband as essential to their own self-realization.

True, not long ago there was a time when work around the house was hard, and it could and did proceed in conjunction with raising children and creating a home. But if modern labor-saving devices are relieving women of the most backbreaking work, they have also done away with the satisfactions it yielded. For girls, too, if machines do it better and faster, it is hard to grow up if there's not enough woman's work to be done. Buying ready-to-wear clothes for her family is a vicarious act. It reflects only her husband's ability to provide the money to buy them, but no unique or essential labors of her own. Since the same is nearly as true for cooking and the home arts, what remain, apart from child-rearing, are the most stultifying tasks—dusting, making beds, washing dishes. And beyond that lie mainly the refinements of homemaking, or what Veblen termed an occupation of ceremonial futility.

Many of the young woman's free-time activities are equally futile. I do not refer only to gardening, which replaces the conspicuous embroidery of an earlier age, or to the bridge circle or country-club life designed to help her husband toward his own type of ceremonial futility. I refer also to much that passes, unexamined, as more valuable pastimes, such as the PTA or the League of Women Voters. When used to cover up a vacuum of truly significant activities, of serious involvement, even these lose the genuine satisfactions they could otherwise confer. For, as Veblen also observed, "Woman is endowed with her share—which there is reason to believe is more than an even share—of the instinct of workmanship, to which futility of life or of expenditure is obnoxious," and such an impulse, when denied expression, leaves them "touched with a sense of grievance too vivid to leave them at rest."

But if a girl tries to fulfill her instinct for workmanship, she is subject to pressures not directed at boys. Many young men show little interest in marriage, even through their early thirties, and are allowed to go their way. At worst, a man may come in for gentle nagging at home, and his friends may tease him about it; but in the final analysis they accept his wish to postpone getting married and founding a family; they tacitly acknowledge that he is not ready yet, that he needs more time to find himself in his work life before he can settle down to family life. Such men are often popular, both in married and unmarried circles, and feel no adverse effect on their sense of accomplishment. In brief, a man is considered a failure if he does not support himself, does not achieve in work, but his marital status little affects people's estimate of him.

All this is very different for the girl. A woman, no matter how gifted or successful in her work life, is judged a failure if she does not marry fairly soon. From adolescence on, therefore, the pressure to marry interferes with her ability to find self-realization in her own personal way. Discrimination usually begins in youth, when there is some indulgence for the boy's nonconformity or revolt because he must "sow his wild oats"; much less tolerance is accorded the girl who seeks to find herself through such a period of nonconformity.

Many a college boy goes through a crisis of identity during his first years away from home, after exposure to many new ideas. Later, in his last years at college, he may suddenly throw himself into his studies, trying to find new identity in his work achievement. Many a girl finds herself in a parallel position; but then she suddenly realizes that with her new dedication to hard work and study, she is failing to compete in the marriage market. Knowing that she wants to have a family one day, and fearing that with her present single-minded absorption it may slip through her fingers, she stops herself dead in her tracks; or worse, she cannot make up her mind which she really wants, and may lose out on both means of self-realization, if marriage has become the only possibility.

Nor is it only the college girl who suddenly kills her excitement about biochemistry because she realizes she is passing up desirable dates. The noncollege girl goes through a similar experience. She too is caught in the realization that society insists she can only find self-realization through an early marriage. So she gives up her tentative new interests as impractical and buckles down to a course in beauty culture or secretarial work. Later, as the young wife of a skilled or unskilled worker, she is exactly as restless and bewildered as the college girl who gave up biochemistry to achieve married life in the suburbs. Neither girl can understand why, though now a success in the eyes of others, all the meaning of life is evading her. This meaning she now looks for in the task of bringing her children up right, which means finding vicarious satisfaction in their lives, with all the consequences discussed earlier.

Here the worker's wife is perhaps the worst off, because she lacks even the secondary gains of her suburban counterpart. Many such women, uprooted too often, no longer try to fill their emptiness with family gossip or church activities, but try to find, if not meaning, at least some escape from emptiness through a job. Unfortunately, it is rarely the kind of work that gives meaning to their lives; but at least it provides association with equals and is preferred to the drudgery of homemaking.

Yet it is not only the instinct for workmanship which is too often frustrated in the modern young woman; frequently it is also her sexual instinct. While sexual difficulties are neither a recent curse of youth, nor restricted to one sex alone, the American attitude toward sex and the educational system have here, too, burdened female youth more than male youth.

Early in this article I referred to the puberty rites of preliterate societies. These elaborate rituals mark the reaching of sexual maturity and assure the initiates of their new adult status. In most of these societies sex was never at all secret nor was pregenital sex experience forbidden. The child learned what sex was all about as he grew up, watching older persons and animals in intercourse. In farming societies, the fecundity of animals is always a central economic issue, so if adult sex is no longer open, at least animal procreation in all its ramifications is still freely observed and discussed in most parts of the world.

Not so in American middle-class society; and what observations are available to the growing child are shrouded in secrecy, if not in embarrassment or outright shame. We all know that shame and embarrassment about the normal bodily functions make sex experiences difficult for much of modern youth. But compensatory efforts to make of sex relations more than they can ever be are equally confounding. I am speaking of the many literary descriptions of intercourse as an earth-shaking event (for example, Hemingway's description in *For Whom the Bell Tolls*). Obviously, one's own sexual experiences, however rewarding, have no such cosmic effect and hence do not seem to come up to par.[8]

Laurence Wylie (p. 248 below) describes how, in a large segment of the French middle classes, the adolescent boy receives his training in love-making from an older woman and then in turn initiates a girl younger than himself in the art he has learned. There is much more than simple experience involved in this way of teaching sex to an inexperienced young man. His very inexperience makes him attractive to the mature woman. In the typical American pattern, ignorance is supposed to be the best teacher of the ignorant in sexual matters. But here the young man's inexperience makes him feel clumsy and insecure in seducing his girl, who, as likely as not, had to seduce him into becoming the seducer in the first place.

American middle-class youth learns about sex in the back seat of a car, or during a slightly drunken party, or because there was nothing better to do to kill boredom (read: sexual frustration and anxiety). The first sexual experience often leaves ineffaceable impressions, marred by a total lack of experience on either side. Both partners feeling anxious and insecure, neither one can offer encouragement to the other, nor can they take comfort from the accomplished sex act, since they cannot be sure that they did it well, all comparisons lacking.

To use Wylie's example again, the young Frenchman not only knows that his inexperience makes him sexually attractive, he also receives the accolade of the person from whom it counts most: an experienced woman has found him not only sexually attractive but from her rich experience (based on comparisons) she has also assured him that he is a manly lover indeed. Thus, secure in his masculinity, he in due course will be sexually attracted by a young girl's inexperience, rather than frightened by it, as his American counterpart usually is. She, feeling that her innocence, or at least her inexperience, makes her attractive, will not feel clumsy because of it, and he, encouraged by previous experience, will feel himself well able to satisfy his girl sexually.[9] Feeling sure that he can satisfy her, she will feel she has satisfied him.

This, of course, is comparing a French "ideal" type of introduction to sex with an American "ideal" type. In actuality, there are as many variations in France as in the United States in the ways youth is introduced to sex. Still, the French way is as typical there as the other is of middle class youth in America. What goes far beyond a mere paradigm of sex behavior in the two countries is the way youth is prepared for the expected sexual role. In France, as in many other lands, both boy and girl from early childhood on are prepared, the first to take the more dominant, the other the more yielding role, not only in sex but also in the family.

In earlier times there have been societies, or at least subsocieties, in which the woman was dominant in the home and even in intimate relations. This situation may still be found in certain segments of the French middle classes, and it used to be characteristic of some orthodox Jewish groups. It seems that successful family life can be organized on such a basis as long as the man's dominance in his sphere is clearly recognized and never challenged by women. The man's sphere is usually the work life, be it in the professions, in business, or politics.

In the orthodox Jewish groups referred to, the man's unquestioned superiority in the all-important religious sphere permitted both to accept gladly the wife's dominance in running the home and often also the shop. With the areas of dominance thus clearly marked out, the wife could be dominant in her sphere without extending it to running her husband's life or her children's. Though such a woman was dominant in the home, no "mom-ism" resulted. Secure in her sphere, it did not occur to her to challenge the man's. More importantly, she did not expect her husband to be dominant at home or in business. Therefore, she was not disappointed in what she expected of him, and hence she did not need to make up for it by nagging him or dominating her children.[10] The woman who engages in "mom-ism" and wishes to "wear the pants" does not act out of an original desire to go her husband one better, but in defense and retaliation.

84

Certainly, our educational system does not prepare the girl to play the more dominant role in the home sphere, nor the more surrendering role, either in sex or other areas of experience. Instead, she is raised in contradiction. On the one hand, she is told that to be feminine means to be yielding, to be courted, and that this is the desirable norm for a woman. She certainly cannot, for example, ask a boy for a date, nor pay the expenses of a date, though in some circles she may sometimes "go dutch."[11]

Contrary to such passivity (waiting to be asked out), where it counts most emotionally, she is taught in school not only to think but also to act for herself. What she is not taught, either at home or in our educational system, is the emotional counterpart of the facts of life: that men and women are neither wholly equal nor by any means opposite sexes, but are complementary; that neither things that are equal nor those that are opposite can be complementary. More importantly, she is not taught wherein men and women are alike—in their talents, aspirations, and emotional needs—and where they are not. From her educationally reinforced but unexamined notion of an equality of the sexes, arise many of the girl's difficulties in her sex relations. For, without clearly understanding her own nature, she does not know where and when to be "feminine" and where and when to be "equal."[12]

For example, in societies in which technology has not yet affected the social conditions of women or their expectations, her sexual life is in far less conflict than in ours. It is still sufficient for her if her lover or husband enjoys sex with her. Since she feels that his enjoyment proves her a good woman, nothing stands in her way of enjoying herself; and, not worrying about whether she is frigid or has an orgastic experience, as likely as not she experiences orgasm. He, not obliged by older tradition or by any newer understanding to provide her with an orgastic experience, can enjoy himself, experience orgasm, and thus help her to experience it herself.

In our own society, the male youth needs as much as ever to have his virility attested by his sexual partner, and the female youth has a parallel need. But by now, the boy also needs to have his girl prove him a man by her so-called "orgastic experience," and the girl is even worse off. She not only has to prove him a man by making him experience orgasm; she must also prove her femininity by the same experience, because otherwise she must fear she is frigid. Sexual intercourse cannot often stand up to such complex emotional demands of proving so many things in addition to being enjoyable.

85

To compound it all, the girl has grown conditioned by all her previous experience in school and college to performing with males on equal grounds, but not on how to complement them.[13] She cannot suddenly learn this in bed. She, trying to make sure that the man has an orgastic experience, and also wondering if she will be able to have one herself, gets so worried that she can truly experience neither, and ends up pretending. In order to prove their manhood or womanhood, the act is now burdened by their having to prove their potency to themselves and to each other, if not also to make the earth shake. Sex becomes another competition of who can make whom have an orgastic experience, and they cannot give up their self-centered needs in the act. The result is that they are unable to enjoy either their mutual desire or the forgetting of self in the experience.

If sexual relations are often less than satisfactory, and if female youth has put work achievement behind her, what is left for the girl by way of self-realization? With home-making now less challenging or satisfying, the children become a concentrated target of the young woman's energies. Here at least, if she starts out feeling less experienced than her mother, she feels considerably more sophisticated. Her world is no longer, as in an older generation, confined to her children, her kitchen, and the church. For years, through the period of her schooling, and perhaps later in a professional occupation, she has worked hard to enlarge her horizon, intellectually and emotionally. Motherhood was depicted to her, and she looked forward to it, as another tremendous, enlarging experience. Yet in reality it forces her to give up most of her old interests, and, unless she is fascinated by the minute developments of the infant, no new and different enrichment is on the horizon. Thus the new world of experience fails to materialize just at the moment when the old enriching experiences are closed, because the infant demands her concentrated attention.

All this is particularly acute with the first child, because the second and third child provide additional content in the mother's life while she cares for the newcomer. I am convinced she will have to find a solution to this problem. It might mean creating something akin to the extended family, through which some societies solved the problem; this meant entrusting part of infant care to the older children or sharing it with relatives. Another solution would be the care of young children by professional people while the mother pursues her individual interests, at least, for part of her time.

In any case, the young mother is now doubly disadvantaged. She cannot find fulfillment in her wifely and motherly role, because she lacks a partner who can complement her in tasks that cannot be mastered alone, or at least not in an emotionally satisfying way. In addition, she is in conflict between her old traditional role and the new image of self-fulfillment through work to which all her schooling has directed her. Or, to put it differently, she is torn between the image of her vicarious role in society as mother and housewife and the self-directed image of herself that developed before marriage. Together, these conflicts are often enough to sour her on motherhood, which she could otherwise fully enjoy.

Resentful in many cases that her husband enjoys what to her seems a fuller life, she tries either to force him to share motherhood with her (which he cannot do without damage to his emotional well-being) or she expects marriage itself to compensate for the frustrated work aspects of her self-realization (which marriage cannot do). Hence she may also sour on her marriage.

The fact that female youth does not react in open conflict with a society that forces her into such an impossible predicament has to do with the actual and socially fostered difference between the sexes. The male delinquent will engage mainly in aggressive acts such as the destruction of property or other forms of aggressive violence; in the "beat," this may be turned inwards, as in the destructive neglect of his own body. In adolescence the female counterpart of violence is sex delinquency, which is less apt to bring the girl into conflict with society.

Yet again this behavior is more often socially imposed than biologically inherent. Many girls who feel "unfeminine" in terms of Hollywood fostered attitudes become sex delinquents to prove they are feminine, or at least to deny that they are not. The nondelinquent, more adult female youth may take things out on her husband and children in less obvious but equally destructive nagging, in a general dissatisfaction that drives her husband to try to achieve for her what only she can achieve for herself.

Others have come to recognize that the problem is largely theirs and are groping for reasonable solutions. Like Negro youth in America, they find themselves a minority with certain psychological advantages. If the majority of young American males can choose their occupations freely but long for more purpose to their labors, the minorities among youth are still fighting with a purpose. Once they do achieve the freedom to pursue self-chosen goals, then the current problems of youth will be no different for boys than for girls, white or Negro.

The elections of 1960 brought an upsurge of purpose, at least in a segment of our population who felt that the new administration had a place for their aspirations, might give them scope to create a better world. This was true for both youth and adult, but youth would not have believed it, if their elders had not shown that their hopes seemed to be justified. As one colleague of mine put it while observing changes on the campus of his university: These students who for years have been rebels without a cause were for the most part too sensible even to rebel. But when they first started to work for Stevenson, and then for Kennedy, and ever since the latter's election, they became dedicated workers for a better future. Yet these, as I say, were only a small segment. The vast majority of the young, like the vast majority of the old, still lacking a direction, seek harder than ever for an empty comfort. This, they hope, may enable them to forget that they have no purpose beyond it, and, as it fails, their search for more of the same becomes increasingly frantic and empty.

Many visitors to Russia, and not only educators, have been struck by how well-behaved Russian children are to their parents, and their parents to them. This mutual respect is entirely different in texture from either fear or the uneasy camaraderie sought by so many parents in our society, where the parent sees his task as that of play companion rather than mentor to the young. Perhaps the explanation lies in what Russian educators have to say about this remarkable difference. They claim that "the good behavior of the children is the result of the clarity and agreement on the part of all teachers [and all adults, we might add] as to their expectations from the children," expectations based on a strong sense of common purpose to create a better social system than has ever existed.[14] Now these for the most part are the children of the people who are still trying to create a new and better society for all, and not of the ruling elite, who already enjoy most of the new advantages. That is why "delinquent" tendencies are found more readily among children of the elite.

Yet even the common purpose of creating a better world, while assuring youth its importance in creating it, is not enough in itself to permit all who seek it to realize themselves. There must be added the clear conception of the usefulness of one's labor; and even then there will always be some who must travel an individual pathway not provided by society, for testing their manhood and worth.

Modern American society has virtually cut off such avenues inside the framework of that society. (Some young people seek them in foreign countries; witness the past attraction of the Spanish Civil War and the current appeal of the Cuban upheaval and of the new African countries.) No longer is there an open frontier for escape from the the oppressive feeling that one cannot prove oneself within a rigidly stratified society. Where is a modern Ishmael to roam? Though he may feel like "methodically knocking off people's hats," he cannot run away to sea because life there is now as regulated and devoid of chances for self-realization as on land. Prevented from knocking people's hats off, or hauled into court or to the psychiatrist for trying to, he feels he has no "substitute for pistol and ball," and may end up using them.

I have said before that for youth to come into his own means to a large degree his replacing the older generation, and that, whether the transition is smooth or hard-won, youth is still on its way. Thus the problem of the generations, when it goes wrong, may be characterized by saying that, whenever the older generation has lost its bearings, the younger generation is lost with it. The positive alternatives of emulation or revolt are then replaced by the lost quality of neither.

And this, I am afraid, is the situation in which large segments of American youth find themselves. They are unhappy when they settle down to continue in a pattern of life that their parents have arranged for them, because they know it to be an empty one. But they find it pointless to rebel, as do those others who, sensing emptiness in the lives prepared for them, fight against it but do not know what to fight for.

Old age is happiest when it can take youth up to the threshold of the good and the new and, like the mythical father of the West, point out the Promised Land to its children, saying: you and only you in a hard fight will have to make this your own; because what is handed down to you, what you have not won for yourselves, is never truly your own.

Youth, on the other hand, is happiest when it feels it is fighting to reach goals that were conceived of but not realized by the generation before them. What the older generation then urgently wished for itself, but had to acknowledge as the hope of the future—this is the legacy of youth. That the preceding generation wished to create such a better world makes it a worthy standard for youth. To come closer to achieving it through its own efforts proves to youth that it is gaining its own rich maturity.

REFERENCES

1. Erik H. Erikson, *Identity and the Life Cycle,* New York: International Universities Press, 1959. Edgar Z. Friedenberg, *The Vanishing Adolescent,* Boston: Beacon Press, 1959. Paul Goodman, *Growing Up Absurd,* New York: Random House, 1960.

2. Bruno Bettelheim, *Symbolic Wounds,* Glencoe, Illinois: The Free Press, 1954.

3. Still, we are not yet too far removed from it in our deeper feelings about what makes for the secure order of things. Witness the feeling of relief, in the last election, when both parties offered young and vigorous candidates for a position being vacated by an old and tired man. Though there is no doubt that millions loved and venerated the old man more than either young candidate, the feeling was prevalent that age had to step aside in favor of youth for the good of the commonweal.

4. Similarly, I know of no Latin scholar who has bought a copy of *Winnie Ille Pooh* for his children; but tens of thousands who never studied Latin have bought the book for their children and have expected them to enjoy it and to acquire an understanding of Latin culture from it. Actually, such a translation is at best a sophomoric prank; taken seriously, it should be obvious on whom the joke is played.

5. J. C. Moloney, "Understanding the Paradox of Japanese Psychoanalysis," *International Journal of Psychoanalysis,* 1953, *34:* 291-303. He also quotes a statement from the *Tokyo Journal of Psychoanalysis* asserting that the task of psychoanalysis is so to strengthen the ego that it can and will respond to the demands of the superego. Those demands in turn are viewed as basically the demands of the emperor, since he represents the all-embracing ethos of his nation.

6. P. Goodman, *op. cit.*

7. A crude index of how far we are from understanding the goals of youth lies in the fact that some United States agricultural extension agents working with 4-H clubs were recently sponsoring a new scheme for keeping youth on the farm: rousing their interest in the care and breeding of race horses!

8. One may also wonder about the success of *Hiroshima, mon Amour* among the intelligentsia. This is a film in which a love affair is significant not because of what it means to the two partners, or of what they find in each other, but because it is played out against a perspective of world history:

the German occupation of France and the bombing of Hiroshima. If the earth has to shake or world history to look on before a love affair can be meaningful for the partners, youth must find itself in awkward straits in its love relations. (I have discussed other difficulties many Americans encounter in finding meaning in their intimate relations in *The Informed Heart,* The Free Press, 1960.)

9. This has been described in Strauss's *Rosenkavalier.*

10. Things changed when such families left the ghettoes and entered modern technological societies. Once the religious sphere lost in importance, while the work sphere grew tremendously important, there were commensurate upheavals in the balance within the family.

11. I can only mention in passing what would call for lengthy discussion: how a society of relative abundance has changed the dating and mating patterns. In other societies it was well known that the greatest attention one could show the courted person was to devote time and attention to him or her. In our society, in which the phrase "time is money" is more than a slogan, money must often make up for time; a boy's car or the money he can spend on his date now replaces the time and attention spent in being with her.

It is not simply that money is made to make up for emotional dedication, which it can never do; it is also made to prove virility, if not even orgastic potency, for which it is equally unsuitable. Many men who doubt their masculinity and virility try to quiet their fears through their social or economic success. Falsely equating virility with beating the other guy in competition, they must come out on top at all costs. But when success is sought not for itself but to make up for something that is missing, it cannot even be enjoyed for what it is, since it cannot make up for what it is not.

The counterpart of such a situation is found in a wife who feels thwarted in her hope for a virile man who can truly make her feel and be a woman. Heedlessly, she spends his money and eggs him on to achieve further status, since these have to make up to her for what they never can, her empty feeling of being a failure as a woman.

But all this comes later in life. It is not yet a problem of youth unless they have been exposed to it in their parents. Similarly, they cannot take their teachers as images to copy, if teachers in their sphere strive for academic success much as members of the business community use money and the status it confers.

12. There is a corresponding confusion in the way we look upon initiative in women, for not all "active" women are unfeminine. Women who strive to "wear the pants" do so for defensive and neurotic reasons, just as the very need to be dominant, whether in man or in woman, is due to thwarted desires, if not also to a thwarted personality. Quite different is the striving to achieve for a purpose, for, like men, some women strive to realize their inner potentials—quite outside any context of competing with other men, other women, or any standard of measurement except their own wish to work toward a purpose. Until we distinguish clearly between the two, we

shall continue to hold back, by labeling "unfeminine," those girls who seek to further their own natural growth and development.

13. Things would not be so bad if the competition were only with persons. To compete with somebody one knows well keeps competition in the human dimension and leads to personal jealousies, hopes and disappointments. But much of the competition where it counts most (merit scholarships, college boards) involves not just a person but rather competing with one's whole age group. It is a competitiveness in the abstract, not against another person, but for a score on a test. As if this were not bad enough, competitions other than for grades and scholastic achievement have entered our educational system. All too often, at the same "educational" place where youngsters compete for grades, they are also competing for dates and a marriage partner.

14. B. Spock, "Russian Children," *Ladies' Home Journal*, October 1960.

TALCOTT PARSONS

Youth in the Context of American Society

THE PASSAGE OF TIME has recently been symbolized by the fact that
we have elected the first President of the United States to be born
in the twentieth century—indeed, well inside it. It is perhaps equally
relevant to remark that we have recently entered an era in which a
substantial proportion of current youth (rather than children) will
experience a major part of their active lives in the twenty-first century. Thus a sixteen-year-old of today will be only fifty-five at the
coming turn of the century.

It is possible that the twentieth century will be characterized by
future historians as one of the centuries of turmoil and transition—
in the modern history of the West, perhaps most analogous to the
seventeenth. It is also likely, however, that it will be judged as one
of the great creative centuries, in which major stages of the process
of building a new society, a new culture, will have occurred. The
tremendous developments in the sciences and in the technologies
deriving from them, the quite new levels of industrialization, and
the spread of the industrial pattern from its places of origin, together with the long series of "emancipations" (e.g., women's suf-
frage and the rapid decline of colonialism) will presumably figure
prominently among its achievements. At the same time, it clearly
has been and will probably continue to be a century of turmoil, not
one of the placid enjoyment of prior accomplishment, but of challenge and danger. It is in this broad perspective that I should like to
sketch some of the problems of American youth, as the heirs of the
next phase of our future, with both its opportunities and its difficulties.

In the course of this century, the United States has emerged at
the forefront of the line of general development, not only because of
its wealth and political power but also—more importantly in the

93

present context—because it displays the type of social organization that belongs to the future. Since during the same period and only a little behind our own stage of progress a somewhat differing and competing version has also emerged in the Communist societies, it is not surprising that there is high tension at both political and ideological levels. Obviously, the meaning of American society presents a world-wide problem, not least to its own citizens and in turn to its younger ones: since they have the longest future ahead of them, they have the most at stake.

Some Salient Characteristics of American Society

Before we take up the specific situation of youth, it will be best to sketch a few of the main features of our society and the ideological discussions about them, with special reference to their effect on youth. The structural characteristic usually emphasized is industrialism. It is certainly true that the United States has developed industrial organization and productivity farther than any other society in history. Not only has it done this on a massive scale, both as regards population and area, but it has also attained by far the highest levels of per-capita productivity yet known. The salience of industrialism in turn emphasizes the economic aspects of social structure: a high evaluation of productivity, the free enterprise system, with the private, profit-oriented business firm as a conspicuous unit of organization, and with private consumption prominent in the disposal of the products of industry. This last feature includes both the high levels of current family income and what may be called the "capitalization" of households through the spread of home ownership, the development of consumer durable goods, and the like.

It would be misleading, however, to overstress this economic aspect. Economic development itself depends on many noneconomic conditions, and economic and noneconomic aspects are subtly interwoven in many ways. The same period (roughly, the present century) which has seen the enormous growth of industrial productivity has also seen a very large relative, as well as absolute, growth in the organization and functions of government. The largest growth of all, of course, is in the armed services, but by no means only there. State and local governments have also expanded. Another prominent development has been that of the legal system, which is interstitial between governmental and nongovernmental sectors of society. I mean here not only legislation and the functioning of courts of law but also the private legal profession, with professional lawyers employed in government in various capacities.

A consideration of the legal profession leads to one of the learned professions in general, the educational organizations in which men are trained, and the cultural systems that form the basis of their competence. The most important development has been the growth of the sciences and their application, not only in industry and the military field but also in many others, notably, that of health. Though they are behind their physical and biological sister disciplines, the sciences dealing with human behavior in society have made very great advances, to an altogether new level. To take only the cruder indices, they have grown enormously in the numbers of trained personnel, in the volume of publications, in the amount of research funds devoted to their pursuit, and the like. All this would not have been possible without a vast expansion of the educational system, relatively greatest at the highest levels. By any quantitative standard, the American population today is by far the most highly educated of any large society known to history—and it is rapidly becoming more so.

Furthermore, this has become in the first instance a society of large organizations, though the tenacious survival of small units (in agriculture, but more broadly in retail trade and various other fields) is a striking fact. (It is important to note that the large organization has many features that are independent of whether it operates in private industry, in government, or in the private nonprofit sector.) It is also a highly urbanized society. Less than ten percent of its labor force is engaged in agriculture, and more than half the population lives in metropolitan areas, urban communities that are rapidly expanding and changing their character.

It is also a society with a great mobility as to persons, place of residence, and social and economic status. It is a society that within about eighty years has assimilated a tremendous number of immigrants, who, though overwhelmingly European in origin, came from a great diversity of national, cultural, and religious backgrounds. Their descendants have increasingly become full Americans, and increasingly widely dispersed in the social structure, including its higher reaches. After all, the current President of the United States is the grandson of Irish immigrants and the first Catholic to occupy that office.

Overriding all these features is the fact that this is a rapidly developing society. There are good reasons for supposing that rapid change is generally a source of unsettlement and confusion, particularly accentuated perhaps if the change is not guided by a set of sharply defined master symbols that tell just what the change is about. The American process of change is of this type; but we can also say that it is not a state of nearly random confusion but in the main is a coherently directional process. Since it is not centrally directed or symbolized, however, it is particularly important to understand its main pattern.

There has been the obvious aspect of growth that is expressed in sheer scale, such as the size of the population, the magnitude and complexity of organization. At the more specifically social levels, however, I should like to stress certain features of the process that may help to make the situation of American youth (as well as other phenomena of our time) more understandable. On the one hand, at the level of the predominant pattern, our value system has remained relatively stable. On the other hand, relative to the value system, there has been a complex process of change, of which structural differentiation is perhaps the most important single feature. It is associated, however, with various others, which I shall call "extending exclusiveness," "normative upgrading," and "an increasing conceptualization of value patterns on the general level." These are all technical terms which, if they are not to be regarded as sociological jargon, need to be elucidated.

Values generally are patterned conceptions of the qualities of meaning of the objects of human experiences; by virtue of these qualities, the objects are considered desirable for the evaluating persons. Among such objects is the type of society considered to be good, not only in some abstract sense but also for "our kind of people" as members of it. The value patterns that play a part in controlling action in a society are in the first instance the conceptions of the good type of society to which the members of that society are committed. Such a pattern exists at a very high level of generality, without any specification of functions, or any level of internal differentiation, or particularities of situation.

In my own work it has proved useful to formulate the dominant American value pattern at this very general level as one of *instrumental activism*. Its cultural grounding lies in moral and (eventually) religious orientations, which in turn derive directly from Puritan traditions. The relevance of the pattern extends through all three of the religious, moral, and societal levels, as well as to others that cannot be detailed here. It is most important to keep them distinct, in particular, the difference between the moral and the societal levels.

In its religious aspect, instrumental activism is based on the pattern Max Weber called "inner-worldly asceticism," the conception of man's role as an instrument of the divine will in building a kingdom of God on earth. Through a series of steps, both in internal cultural development and in institutionalization (which cannot be detailed here), this has produced a conception of the human condition in which the individual is committed to maximal effort in the interest of valued *achievement* under a system of normative order. This system is in the first instance moral, but also, at the societal level, it is embodied in legal norms. Achievement is conceived in "rational" terms, which include the maximal objective understanding of the empirical conditions of action, as well as the faithful adherence to normative commitments. It is of great importance that, once institutionalized, the fulfillment of such a value pattern need not be motivated by an explicit recognition of its religious groundings.

One way of describing the pattern in its moral aspects is to say that it is fundamentally individualistic. It tends to maximize the desirability of autonomy and responsibility in the individual. Yet this is an institutionalized individualism, in that it is normatively controlled at the moral level in two ways. First, it is premised on the conception of human existence as serving ends or functions beyond those of physical longevity, or health, or the satisfaction of the psychological needs of the personality apart from these value commitments. In a sense, it is the building of the "good life," not only for the particular individual but also for all mankind—a life that is accounted as desirable, not merely desired. This includes commitment to a good society. Second, to implement these moral premises, it is necessary for the autonomous and responsible achievements of the individual to be regulated by a normative order—at this level, a moral law that defines the relations of various contributions and the patterns of distributive justice.

The society, then, has a dual meaning, from this moral point of view. On the one hand, it is perhaps the primary field in which valued achievement is possible for the individual. In so far as it facilitates such achievements, the society is a good one. On the other hand, the building of the good society (that is, its progressive improvement) is the primary goal of valued action—along with such cultural developments as are intimately involved in social progress,

97

such as science. To the individual, therefore, the most important goal to which he can orient himself is a contribution to the good society.

The value pattern I am outlining is activistic, therefore, in that it is oriented toward control or mastery of the human condition, as judged by moral standards. It is not a doctrine of passive adjustment to conditions, but one of active adaptation. On the other hand, it is instrumental with reference to the source of moral legitimation, in the sense that human achievement is not conceived as an end in itself but as a means to goals beyond the process and its immediate outcome.

This value pattern implies that the society is meant to be a developing, evolving entity. It is meant to develop in the direction of progressive "improvement." But this development is to be through the autonomous initiative and achievements of its units—in the last analysis, individual persons. It is therefore a society which places heavy responsibilities (in the form of expectations) on its individual members. At the same time, it subjects them to two very crucial sets of limitations which have an important bearing on the problem of youth.

One of these concerns the "moralism" of the value system—the fact that individualism is bound within a strongly emphasized framework of normative order. The achievement, the success, of the individual must ideally be in accord with the rules, above all, with those which guarantee opportunity to all, and which keep the system in line with its remoter values. Of course, the more complex the society, the greater the difficulty of defining the requisite norms, a difficulty which is greatly compounded by rapid change. Furthermore, in the interest of effectiveness, achievement must often be in the context of the collective organization, thus further limiting autonomy.

The second and for present purposes an even more crucial limitation is that it is in the nature of such a system that it is not characterized by a single, simple, paramount goal for the society as a system. The values legitimize a *direction* of change, not a terminal state. Furthermore, only in the most general sense is this direction "officially" defined, with respect to such famous formulae as liberty, democracy, general welfare, and distributive justice. The individual is left with a great deal of responsibility, not only for achieving *within* the institutionalized normative order, but for his own interpretation of its meaning and of his obligations in and to it.

Space forbids detailing the ramifications of this value system. Instead, it is necessary to my analysis to outline briefly the main features of the process of social change mentioned above. The suggestion is that the main pattern of values has been and probably will continue to be stable, but that the structure of the society, including its subsystem values at lower levels, has in the nature of the case been involved in a rapid and far-reaching process of change. This centers on the process of differentiation, but very importantly it also involves what we have referred to as inclusion, upgrading, and increasing generalization. I shall confine my discussion here to the structure of the society, though this in turn is intimately connected with problems concerned with the personality of the individual, including his personal values.

Differentiation refers to the process by which simple structures are divided into functionally differing components, these components becoming relatively independent of one another, and then recombined into more complex structures in which the functions of the differentiated units are complementary. A key example in the development of industrial society everywhere is the differentiation, at the collectivity level, of the unit of economic production from the kinship household. Obviously, in peasant economies, production is carried out by and in the household. The development of employing organizations which are structurally distinct from any household is the key new structural element. This clearly means a loss of function to the old undifferentiated unit, but also a gain in autonomy, though this in turn involves a new dependency, because the household can no longer be self-subsistent. The classical formula is that the productive services of certain members (usually the adult males) have been alienated from the organization directly responsible for subsistence and thus lost to the household, which then depends on money income from occupational earnings and in turn on the markets for consumers' goods.

These losses, however, are not without their compensations: the gain in the productivity of the economy and in the standard of living of the household. This familiar paradigm has to be generalized so as to divest it of its exclusively economic features and show it as the primary characterization of a very general process of social change. First, it is essential to point out that it always operates simultaneously in both collectivities and individual roles. Thus, in the example just given, a new type of productive organization which is not a household or (on more complex levels) even a family farm has

99

to be developed. The local community no longer consists only of farm households but also of nonproducing households and productive units—e.g., firms. Then the same individual (the head of the household) has a dual role as head of the family and as employee in a producing unit (the case of the individual enterpreneur is a somewhat special one).

By the extension of inclusiveness, I mean that, once a step of differentiation has been established, there is a tendency to extend the new pattern to increasing proportions of the relevant population of units. In the illustrative case, the overwhelming tendency that has operated for well over a century has been to reduce the proportion of households which are even in part economically self-sufficient, in the sense of a family farm, in favor of those whose members are gainfully employed outside the household. This is a principal aspect of the spread of industrialization and urbanization. The same logic applies to newly established educational standards, e.g., the expectation that a secondary-school education will be normal for the whole age cohort.

Normative upgrading means a type of change in the normative order, to which the operation of units, both individual and collective, is subject. It is a shift from the prescription of rules by a special class or unit in a special situation to more generalized norms having to do with more inclusive classes of units in wider ranges of situations. Thus the law that specifies that a railway engine must be equipped with a steam whistle to give warning at crossings has by court interpretation been generalized to include any effective warning signal (since oil-burning locomotives are not equipped with steam).[1] But in a sense parallel to that in which differentiation leads to alienation from the older unit, normative upgrading means that the unit is left with a problem, since the rules no longer give such concretely unequivocal guidance to what is expected. If the rule is general enough, its application to a particular situation requires interpretation. Such upgrading, we contend, is a necessary concomitant of the process of differentiation.

When we speak of norms, we mean rules applying to particular categories of units in a system, operating in particular types of situations. For example, individual adults may not be employed under conditions which infringe on certain basic freedoms of the individual. The repercussions of a step in differentiation, however, cannot be confined to this level; they must also involve some part of the value system; this is to say, the functions of the differentiated categories of units, which are now different from one another, must not

100

only be regulated but also legitimized. To use our example again, it cannot be true that the whole duty of the fathers of families is to gain subsistence for their households through making the household itself productive, but it becomes legitimate to support the household by earning a money income through work for an outside employer and among other things to be absent from the household many hours a week. At the collectivity level, therefore, a business that is not the direct support of a household (such as farming) must be a legitimate way of life—that is, the unit that employs labor for such purposes, without itself being a household, must be legitimate. This requires defining the values in terms sufficiently general to include both the old and the new way of life.

The values must therefore legitimize a structural complex by which economic production and the consumption needs of households are met simultaneously—that is, both the labor markets and the markets for consumers' goods. For example, this structural complex is of focal importance in the modern (as distinguished from medieval) urban community. The value attitude that regards the rural or the handicraft way of life as morally superior to the modern urban and—if you will—industrial way (a common attitude in the Western world of today) is an example of the failure of the adequate value generalization that is an essential part of institutionalizing the process of structural change.

To sum up, we may state that both the nature of the American value pattern and the nature of the process of change going on in the society make for considerable difficulties in the personal adjustment of individuals. On the one hand, our type of activism, with its individualistic emphases, puts a heavy responsibility for autonomous achievement on the individual. On the other hand, it subjects him to important limitations: he must not only be regulated by norms and the necessity of working cooperatively, in collective contexts; he must also interpret his own responsibilities and the rules to which he is subject. Beyond that, ours is a society which in the nature of its values cannot have a single clear-cut societal goal which can be dramatically symbolized. The individual is relegated to contributions which are relatively specialized, and it is not always easy to see their bearing on the larger whole. Furthermore, the general erosion of traditional culture and symbols, which is inseparable from a scientific age, makes inadequate many of the old formulae once used to give meaning and legitimation to our values and achievements. This is perhaps true in particular of the older religious grounding of our values.

Not unrelated to these considerations is the very fact of the *relative* success of the society in developing in relation to its values. Not only is there a high general standard of living, which, it should be remembered, means the availability of facilities for *whatever* uses are valued; e.g., increased income may allow for attending prize fights or symphony concerts—a not inconsiderable amount has been going into the latter channel. There is certainly a much better standard of minimum welfare and general distributive justice now than in our past. However much remains to be done, and it is clearly considerable, it is no longer possible to contend that poverty, misery, preventable illness, etc., are the primary lot of the average American. Indeed, the accent has shifted to our duty to the less favored portions of the world. Furthermore, for the average individual, it is probable that opportunity is more widely open than in any large-scale society in history to secure education, access to historically validated cultural goods, and the like. But perhaps it can be seen that, in the light of this all too brief analysis, the great problem has come to be, what to do with all these advantages—not, as has so often been true, how to avoid the worst disasters and take a few modest little steps forward.[2] To be sure, there is a very real danger of the collapse of all civilization through nuclear war; but somehow that danger fails to deter people from making significant investments in the future, not only for themselves as individuals, but also for the society as a whole.

The Position of American Youth

It is in this broad picture of the American social structure and its development that I should like to consider the position of American youth. Contrary to prevalent views that mainly stress the rising standard of living and the allegedly indulgent and easy life, I think it is legitimate to infer that the general trend of development of the society has been and will continue to be one which, by and large, puts greater rather than diminished demands on its average individual citizen—with some conspicuous exceptions. He must operate in more complex situations than before. He attempts to do many things his predecessors never attempted, that indeed were beyond their capacities. To succeed in what he attempts, he has to exercise progressively higher levels of competence and responsibility. These inferences seem to me inescapable when full account of the nature of the society and its main trends of development is taken.

If capacities and relevant opportunities developed as rapidly as do demands, it would follow that life on the average would be neither more nor less difficult. There seems reason to believe that if anything demands have tended somewhat to outrun the development of capacities—especially those for orienting to normatively complex situations—and in some respects even opportunities, and that this is a major source of the current unrest and malaise. My broad contention, taking due account of the process of change just outlined, is that this society, however, is one that is relatively well organized and integrated with reference to its major values and its major trends of development. If those values are intact and are by and large shared by the younger generation (there seems to be every indication that they are), then it ought to be a society in which they can look forward to a good life. In so far as their mood is one of bewilderment, frustration, or whatever, one should look for relatively specific sources of difficulty rather than to a generalized malintegration of the society as a whole.

It may be well to set the tone of the following analysis by an example of the ways in which current common sense can often misinterpret phenomena that raise distressing problems. American society, of course, is known for its high divorce rate. Until the peak following World War II, moreover, the trend was upward throughout the century, though since then it has appreciably declined. This divorce rate has widely been interpreted as an index of the "disintegration of the family" and, more importantly, of the levels of moral responsibility of married persons.

That it results in increased numbers of broken families is of course true, though the seriousness of this is mitigated by the fact that most divorces occur between childless couples and that most divorced persons remarry, a large proportion stably. In any case, the proportion of the population of marriageable age that is married and living with their spouses is now the highest it has been in the history of the census.

The main point, however, is that this point of view fails to take into account the increased strain put on the marriage relationship in the modern situation. In effect, it says, since an increased proportion fail in a difficult task relative to those who previously failed in an easier task, this increased rate of failures is an index of a declining level of responsibility; seen in this light, this interpretation is palpably absurd, but if the underlying situation is not analyzed, it is plausible.

The increased difficulty of the task has two main aspects. One is the increased differentiation of the nuclear family from other structures in which it was formerly embedded, notably the farm and other household or family enterprises from which economic support was derived. This differentiation deprives the family and the marriage relationship within it of certain bases of structural support. This is clearly related to the component of freedom mentioned above; the freedom of choice of marriage partners is clearly related to the spread of the view that really serious incompatibility may justify breaking the marriage tie.

The other factor is the enhanced level of expectation in functioning outside the family for both adults and children. For adults, particularly men, the central obligation concerns the levels of responsibility and competence required by their jobs; for children, these requirements of growing up in a more complex and competitive world, going farther in education, and undertaking substantially more autonomous responsibility along the way impose greater demands than before. It is my impression that the cases in which marriage was undertaken irresponsibly are no more numerous than in any other time, and that divorce is not often lightly resorted to but is a confession of failure in an undertaking in which both parties have usually tried very hard to succeed.[3]

I cite this example because it is a conspicuous special case of the more general considerations I wish to discuss. The first keynote here is the rising general level of expectations. The primary reference point, of course, is that of adult roles at their peak of responsibility in middle age. The most prominent example is that of the higher levels of masculine occupational roles, in which (in those with technical emphasis) the requisite levels of training and technical competence are continually rising. With respect to managerial roles, the size and complexity of organizations is increasing, and hence the requirements necessary for their successful management also. Similar things, however, are true in various other fields. Thus the whole range of associational affairs requires membership support for leadership as well as responsible leadership itself, both of which involve complicated responsibilities. These range from the many private associations and "good causes" through participation on boards and staffs (including university departments and faculties) to participation through voting and other forms of exercising public responsibility.

The family in this context is a further case. The feminine role is typically anchored in the first instance in the family. Family duties may not be more onerous in such senses as drudgery and hard work than they were, but they involve a higher level of competence and responsibility, particularly, though not exclusively, in the field of the psychological management of both children and husbands, as well as of selves—the latter because wives are now far more autonomous on the average than they were. What we may call the independence training of children is more delicate and difficult than was the older type of training in strict obedience—that is, if autonomy for the young is to be accompanied by high levels of self-discipline and responsibility. But in addition, the typical married woman participates far more extensively outside the home than she formerly did, and in particular she forms a rapidly increasing proportion in the labor force.

Perhaps the central repercussion of this general upgrading of expectations (and hence of the norms with which conformity is expected) on the situation of youth is in the field of formal education. Here, of course, there has been a steady process of lengthening the average period of schooling, with the minimum satisfactory norm for all approaching the completion of high school, while nearly forty percent of the total age cohort now enter college, and a steadily increasing percentage complete college. Finally, by far the most rapidly growing sector has been that of postgraduate professional education. Uneven as standards are, and unsatisfactory as they are at many points, there is no solid evidence of a general tendency to deterioration and much evidence of their improvement, especially in the best schools at all levels.[4]

It seems fair, then, to conclude that in getting a formal education the average young American is undertaking a more difficult, and certainly a longer, job than his father or mother did, and that it is very likely that he is working harder at it. A growing proportion is prolonging formal education into the early adult years, thus raising important problems about marriage, financial independence, and various other considerations.

Furthermore, he is doing this in a context in which, both within and outside the school, he must assume more autonomous responsibility than did his predecessors. In the school itself—and in college —the slow though gradual trend has been in the direction of a mildly "progressive" type of education, with a diminution of the amount of drill and learning by rote. In certain respects, parents have grown distinctly more permissive within the family and with regard to their children's activities outside. This throws an important stress on the child's relations to his age peers, one that becomes particularly important in adolescence. This is the area least under adult control, in which deviant tendencies can most readily be mutually reinforced, without being immediately checked by adult intervention. This is to say that in general the educational process puts increased demands on the younger group.

Three other factors seem involved in this situation of strain from the combination of enhanced expectations and autonomy. They concern one aspect of the psychological preparation for the tasks of maturing, one aspect of the choices that are open, and one aspect of the situation with reference to normative regulation.

First, with respect to psychological preparation, there seems to have been a trend within the family to *increase* the dependency of the young pre-oedipal child, particularly on the mother, of course. This trend is the consequence of the structural isolation of the nuclear family. There is less likelihood of there being close relatives either directly in the home or having very intensive and continual contact with the family. For middle-class families, the virtual disappearance of the domestic servant has also left less room for a division of responsibility for child care. Further, the proportion of very large families with five or more children has been sharply decreasing, while those with three and four children have been increasing. All these factors contribute to a concentration of relationships within the family and of the parents' (especially the mother's) sanctioning powers—both disciplinary and rewarding.

Psychological theory, however, indicates that under the proper circumstances this enhanced dependency contributes to developing motivations for high levels of achievement. These circumstances include high levels of aspiration for the child on the part of the parents and the use of the proper types of discipline. The essential point is that high dependency provides a very strong motivation to please the parent. This in turn can be used to incite him to learn what the parent sets him, if he is suitably rewarded by parental approval. The general findings of studies on the types of discipline used in middle-class families—the use of the withdrawal of love and approval as the predominant type of negative sanction—seem to fit in this picture.

The dependency components of motivation, however, are seldom if ever fully extinguished. The balance is so delicate in their relation to the autonomous components that it is easily upset, and in many cases this is a source of considerable strain. Attempting to maintain this balance, for example, may very well contribute to the great increase in the practice of "going steady" and its relation to the trend to early marriages. Emerging in adolescence, the dyadic heterosexual relation is the main component of the relational system of youth that articulates most directly with the earlier dependency complex—though some of it may also be expressed in same-sex peer groups, and indeed in "crushes" on the teacher. It is striking that the main trend seems to be toward intensive, and not merely erotic but diffuse, dyadic relations, rather than to sexual libertinism. This is in turn reflected in the emotional intensity of the marriage relationship and hence in the elements of potential strain underlying the problem of divorce.

This brings me to the second of the factors mentioned above, the range of choices open. A progressive increase in this range is a consequence of the general process of social change sketched above, namely, differentiation in the structure of the society. As this process goes on, types of interest, motivation, and evaluation that were embedded in a less differentiated complex come to be separated out, to become more autonomous and more visible in that they are freed from more ascriptive types of control. Ties to class and family, to local community and region become more flexible and hence often "expendable" as more choices become available.

One of the most conspicuous cases in relation to the present interest is the erotic component of sex relations. In an earlier phase of our society, it was rather rigidly controlled even within marriage, indeed, not infrequently it was partially suppressed. The process by which it has become differentiated, allowing much greater freedom in this area, is closely related to the differentiation of function and the structural isolation of the nuclear family.[5] In a society in which autonomous freedom is so widespread, there is much greater freedom in this field as in many others, not only in practice but also in portrayals on the stage, in the movies and television, and in the press, magazines, and books.

In this connection, since much of the newer freedom is illegitimate in relation to the older standards (normative upgrading and value generalization take time), it is very difficult to draw lines between the areas of new freedom in process of being legitimated and the types which are sufficiently dysfunctional, even in the new state of society, so that the probability is they will be controlled or even suppressed. The adolescent in our society is faced with difficult

107

problems of choice and evaluation in areas such as this, because an adequate codification of the norms governing many of these newly emancipated areas has not yet been developed.

The third factor, that of normative regulations, is essentially a generalization of the second factor. We have maintained (though of course without documentation) that, contrary to various current opinions, the basic pattern of American values has not changed. Value patterns, however, are only part of the normative culture of the society. At the lower levels, both at the more specific levels of values and of what we technically call norms, it is in the nature of the type of process of change we have been discussing that there should be a continual reorganization of the normative system. Unfortunately, this does not occur as an instantaneous adjustment to the major innovations, but is a slow, uneven, and often painful process. In its course, at any one time (as we have noted), there are important elements of indeterminacy in the structure of expectations—not simply in the sense that there are areas of freedom in which autonomous decision is expected, but also in the sense that, where people feel there ought to be guidance, it is either lacking altogether, or the individual is subject to conflicting expectations that are impossible to fulfill all at once. This is the condition that some sociologists, following Durkheim, call *anomie*.

There seems to be an important reason why this source of strain and disturbance bears rather more heavily on the younger generation than on others. This is owing to the fact that the major agents for initiating processes of change lie in other sectors of the society, above all, in large-scale organization, in the developments of science and technology, in the higher political processes, and in the higher ranges of culture. Their impact tends to spread, and there is a time lag in change between the locations of primary change and the other parts of the social structure.

Though there is of course much unevenness, it seems correct to say that, with one major exception, the social structures bearing most directly on youth are likely to be rather far down the line in the propagation of the effects of change. These are the family and the school, and they are anchored in the local residential community. The major exception is the college, and still more, the university, which is one of the major loci of innovation and which can involve its students in the process more directly.

By and large, it seems fair to suggest that adults are on the average probably more conservative in their parental roles than when their children are not involved, and that this is typical of most of their roles outside the family. Similarly, schools, especially elementary and secondary schools, are on the whole probably more conservative in most respects than are the organizations that employ the fathers of their children. In the present phase of social development, another important institution of the residential community, the parish church or synagogue, is probably distinctly on the conservative side as a rule.

This would suggest that, partly as a matter of generation lag, partly for more complex reasons of the sort indicated, the adult agencies on which the youth most depends tend to some extent to be "out of tune" with what he senses to be the most advanced developments of the time. He senses that he is put in an unfair dilemma by having to be so subject to their control.

If we are right in thinking that special pressures operate on the younger generation relative to the general pressures generated by social change, on the other side of the relationship there are factors which make for special sensitivities on their part. The residua of early dependency, as pointed out above, constitute one such factor. In addition, the impact on youth of the general process of social differentiation makes for greater differences between their position and that of children, on the one hand, and that of adults, on the other, than is true in less differentiated societies. Compared to our own past or to most other societies, there is a more pronounced, and above all (as noted) an increasingly long segregation of the younger groups, centered above all on the system of formal education. It may be argued especially that the impact of this process is particularly pronounced at the upper fringe of the youth period, for the rapidly increasing proportion of the age cohort engaged in higher education—in college, and, very importantly, in postgraduate work. These are people who are adults in all respects except for the element of dependency, since they have not yet attained full occupational independence.

The Youth Culture

The question may now be raised as to how young people react to this type of situation. Obviously, it is a highly variegated one and therefore occasions much diversity of behavior, but there are certain broad patterns which can be distinguished. These may be summed up under the conception, now familiar to social scientists, of a rela-

109

tively differentiated "youth culture." Perhaps S. N. Eisenstadt is its most comprehensive student, certainly in its comparative perspective.[6]

It is Eisenstadt's contention that a distinctive pattern of values, relationships, and behavior for youth tends to appear and become more or less institutionalized in societies that develop a highly universalistic pattern of organization at the levels of adult role involvements. Since all lives start in the family, which is a highly particularistic type of structure, there is not only the difficulty of rising to higher levels within the same type of relationship system, but also of learning to adjust to a very different type. What has been discussed above as the enhancement of dependency in early childhood is a special case of this general proposition. Totalitarian societies attempt to bring this period under stringent centralized control through officially organized, adult-directed youth organizations such as the Soviet *Komsomols*, or earlier, the *Hitlerjugend*. In democratic societies, however, it tends to be relatively free, though in our own it is rather closely articulated with the system of formal education through a ramifying network of extracurricular activities.

As a consequence of youth's being exposed to such strains, it might be expected that youth culture would manifest signs of internal conflict and that it would incorporate elements of conformity as well as of alienation and revolt. In nonrational, psychological terms, rather than in terms of rational aims, youth culture attempts to balance its need for conforming to the expectations of the adult agencies most directly involved (parents and the local residential community) with some kind of outlet for tension and revolt and with some sensitivity to the winds of change above and beyond its local situation.

For two reasons, one would expect to find the fullest expression of these trends at the level of the peer group. For one thing, this group is the area of greatest immunity to adult control; indeed, the range of its freedom in this respect is particularly conspicuous in the American case. The other reason is that this is the area to which it is easiest to displace the elements of dependency generated in early experience in the family—on the one hand, because the strong stress on autonomy precludes maintaining too great an overt dependence on parents or other adult agencies, and, on the other, because the competitive discipline of school achievement enforces autonomous responsibility in this area. The peer group then gradually differentiates into two components, one focusing on the cross-sex relationship and one focusing on "activities," some of which occur within the one-sex group, others, relatively nonerotic, in mixed groups.

In general, the most conspicuous feature of the youth peer group is a duality of orientation. On the one hand, there tends to be a compulsive independence in relation to certain adult expectations, a touchy sensitivity to control, which in certain cases is expressed in overt defiance. On the other hand, within the group, there tends to be a fiercely compulsive conformity, a sharp loyalty to the group, an insistence on the literal observance of its norms, and punishment of deviance. Along with this goes a strong romantic streak. This has been most conspicuous in the romantic love theme in the cross-sex relationship, but it is also more generalized, extending to youth-culture heroes such as athletes and group leaders of various sorts, and sometimes to objects of interest outside the youth situation.

It is my impression (not easy to document) that important shifts of emphasis in American youth culture have occurred in the last generation. For the main trend, notably the increasingly broad band we think of as middle class, there has been a considerable relaxation of tension in both the two essential reference directions, toward parents and toward school expectations—though this relaxation is distinctly uneven. In the case of the school, there is a markedly greater acceptance of the evaluation of good school work and its importance for the future. This, of course, is associated with the general process of educational upgrading, particularly with the competition to enter good colleges and, at the next level, especially for students at the better colleges, to be admitted to graduate schools. The essential point, however, is that this increased pressure has been largely met with a positive response rather than with rebellion or passive withdrawal. The main exception is in the lowest sector, where the pattern of delinquency is most prominent and truancy a major feature. This is partly understandable as a direct consequence of the upgrading of educational expectations, because it puts an increased pressure on those who are disadvantaged by a combination of low ability, a nonsupportive family or ethnic background.

As to youth's relation to the family, it seems probable that the institutionalizing of increased permissiveness for and understanding of youth-culture activities is a major factor. The newer generation of parents is more firmly committed to a policy of training serious independence. It tolerates more freedom, and it expects higher levels of performance and responsibility. Further, it is probably true that the development of the pattern of "going steady" has drained off some tension into semi-institutionalized channels—tension formerly expressed in wilder patterns of sexual behavior. To be sure, this creates a good many problems, not only as to how far the partners will go in their own erotic relations, but also possibly premature commitments affecting future marriage. It may be that the pendulum has swung too far and that adjustments are to be expected.

Within this broad framework, the question of the content of peer-group interests is important. What I have called the romantic trend can be broadly expressed in two directions; the tentative terms "regressive" and "progressive" are appropriate, if not taken too literally. Both components are normally involved in such a situation, but their proportions and content may vary. They derive specifically from the general paradigm of social change outlined above, the former, at social levels, tending to resist change, the latter to anticipate and promote it.

One of the most striking interests of American youth culture has been in masculine physical prowess, expressed in particular in athletics. It seems quite clear that there has been a declining curve in this respect, most conspicuous in the more elite schools and colleges, but on the whole it is a very general one, except for the cult of violence in the delinquent sector. The cult of physical prowess has clearly been a reflex of the pressure to occupational achievement in a society in which brains rather than brawn come increasingly to count. From this point of view, it is a regressive phenomenon.

The indication is that the lessened concentration on this cult is an index of greater acceptance of the general developmental trend. Alcohol and sex are both in a somewhat different category. For the individual, they are fields of emancipation from the restrictions of childhood, but they are definitely and primarily regressive in their significance for the adult personality. However, as noted above, the emancipation of youth in this respect has been connected with a general emancipation which is part of the process of differentiation in the adult society, which permits greater expressiveness in these areas. I have the impression that a significant change has occurred from the somewhat frenetic atmosphere of the "flaming youth" of the 1920's and to some extent of the 1930's. There is less rebellion in both respects, more moderation in the use of alcohol, and more "seriousness" in the field of sexual relations. Youth has become better integrated in the general culture.

On the other side, the progressive one, the most important phenomena are most conspicuous at the upper end of the range, both in
terms of the sociocultural level and of the stage of the life cycle.
This is the enormous development of serious cultural interests among
students in the more elite colleges. The most important field of these
interests seem to be that of the arts, including highbrow music,
literature, drama, and painting.

The first essential point here is that this constitutes a very definite
upgrading of cultural standards, compared with the philistinism of
the most nearly corresponding circles in an earlier generation. Second, however, it is at least variant and selective (though not, I
think, deviant) with respect to the main trends of the society,
since the main developments in the latter are on the "instrumental"
rather than the "expressive" side. As to the special involvement of
elite youth in the arts, it may be said that youth has tended to become a kind of "loyal opposition" to the main trends of the culture,
making a bid for leadership in a sphere important to a balanced society yet somewhat neglected by the principal innovating agencies.

The question of youth's relation to the political situation is of
rather special interest and considerable complexity. The suscepti-
bility of youth groups to radical political ideologies, both left and
right, has often been remarked. It appears, however, that this is a
widely variant phenomenon. It seems to be most conspicuous, on
the one hand, in societies just entering a more "developed" state, in
which intellectuals play a special role and in which students, as potential intellectuals, are specially placed. In a second type of case,
major political transitions and instabilities are prominent, as in several European countries during this century, notably Germany.

Seen in this context, American youth has seemed to be apathetic
politically. During the 1930's and 1940's, there was a certain amount
of leftist activity, including a small Communist contingent, but the
main trend has certainly been one of limited involvement. Recently,
there seems to have been a kind of resurgence of political interest
and activity. It has not, however, taken the form of any explicit,
generalized, ideological commitment. Rather, it has tended to focus
on specific issues in which moral problems are sharply defined,
notably in race relations and the problems of nuclear war. It does
not seem too much to say that the main trend has been in accord
with the general political characteristics of the society, which has
been a relatively stable system with a strong pluralistic character.
The concomitant skepticism as to generalized ideological formulae

113

is usually thought deplorable by the moralists among our intellectuals. In this broad respect, however, the main orientation of youth seems to be in tune with the society in which they are learning to take their places.

The elements in youth culture that express strain because of deviations from the main standards of the adult society are by no means absent. One such deviation is what we have called the "romantic," the devotion to expectations unrealistically simplified and idealized with respect to actual situations. A particularly clear example has been the romantic love complex. It is interesting, therefore, that a comparable pattern seems to have appeared recently in the political field, one that is connected with a pervasive theme of concern: the "meaningfulness" of current and future roles in modern industrial society.

As is brought out in Kenneth Keniston's contribution to this issue, in the field of politics, one not very explicit interpretation of a meaningful role for youth in general is to exert a major personal influence on determining the "big" political decisions of our time. The realistic problem, of course, is the organization of large-scale societies on bases that are not rigidly fixed in tradition, not authoritarian, and not unduly unstable. In this respect, public opinion (though in the long run extremely important) is necessarily diffuse and, with few exceptions, unable to dictate particular decisions. The main policy-making function is of necessity confined to relatively few and is the special responsibility of elected representatives who, in large-scale societies, become professionalized to a considerable degree. The average adult citizen, even if high in competence and responsibility, is excluded from these few. Yet this is not to say that in his role as citizen his responsibilities are meaningless or that his life in general can become meaningful only if his principal concerns (e.g., his non-political job) are sacrificed to the attempt to become a top "influential" in national politics. If this were true, representative democracy as we know it would itself be meaningless. The alternative, however (if large-scale society is to exist at all), is not populistic direct democracy but dictatorship.

This particular syndrome, of course, is a part of a larger one: the general difficulty of accepting the constraints inherent in large-scale organizations—in particular, the "instrumental" aspect of roles other than those at the highest levels. We have already pointed out some of the features of our developing social system that make this a focus of strain. Equally, through the development of institutionalized individualism, there is a whole series of factors making for an increasing rather than a diminishing autonomy. The question, however, concerns the spheres in which the autonomy of various categories of individuals can operate. Differentiation inevitably entails mutual dependence: the more differentiation, the more dependence. In a system characterized by high levels of differentiation, it is to be expected that organizational policy making will also become differentiated. Hence, only a few will become very intimately concerned with it. The problem of what mechanism can control these few is indeed a complex one which cannot be analyzed here. The political role, however, seems to provide particularly striking evidence of a romantic element in current youth ideology.

Perhaps the most significant fact about current youth culture is its concern with meaningfulness. This preoccupation definitely lies on the serious and progressive side of the division I have outlined. Furthermore, it represents a rise in the level of concern from the earlier preoccupation with social justice—even though the problem of race relations is understandably a prominent one. Another prominent example is the much discussed concern with problems of "identity." This is wholly natural and to be expected in the light of *anomie*. In a society that is changing as rapidly as ours and in which there is so much mobility of status, it is only natural that the older generation cannot provide direct guidance and role models that would present the young person with a neatly structured definition of the situation. Rather, he must find his own way, because he is pushed out of the nest and expected to fly. Even the nature of the medium in which he is to fly is continually changing, so that, when he enters college, there are many uncertainties about the nature of opportunities in his chosen field on completing graduate school. His elders simply do not have the knowledge to guide him in detail.

It is highly significant that the primary concern has been shifting since early in the century from the field of social justice to that of meaningfulness, as exemplified by the problem of identity—except for the status of special groups such as the Negro. In terms of the social structure, this enhances the problem of integration, and focuses concern more on problems of meaning than on those of situation and opportunity in the simpler sense. It is a consequence of the process of social change we have outlined.

It is also understandable and significant that the components of anxiety that inevitably characterize this type of strained situation should find appropriate fields of displacement in the very serious,

115

real dangers of the modern world, particularly those of war. It may also be suggested that the elite youth's resonance to the diagnosis of the current social situation in terms of conformity and mass culture should be expected.[7] Essentially, this diagnosis is an easy disparagement of the society, which youth can consider to be the source of difficulty and (so it seems to them) partially unmanageable problems.

Conclusion

The above analysis suggests in the main that contemporary American society is of a type in which one would expect the situation of youth to involve (certainly, by the standards of the society from which it is emerging) rather special conditions of strain. As part of the more general process of differentiation to which we have alluded, youth groups themselves are coming to occupy an increasingly differentiated position, most conspicuously, in the field of formal education. Though an expanding educational system is vital in preparing for future function, it has the effect of segregating (more sharply and extensively than ever before) an increasing proportion of the younger age groups. The extension of education to increasingly older age levels is a striking example.

The other main focus of strain is the impact on youth of the pace and nature of the general process of social change. This is especially observable in the problem of *anomie*. In view of this change, youth's expectations cannot be defined either very early or very precisely, and this results in considerable insecurity. Indeed, the situation is such that a marked degree of legitimate grievance is inevitable. Every young person is entitled in some respects to complain that he has been brought into "a world I never made."

To assess the situation of American youth within the present frame of reference presents an especially difficult problem of balance. This is an era that lays great stress, both internally and externally, on the urgencies of the times, precisely in the more sensitive and responsible quarters. Such a temper highlights what is felt to be wrong and emphasizes the need for change through active intervention. With reference to the actual state of society, therefore, the tendency is to lean toward a negative evaluation of the status quo, because both the concrete deficiencies and the obstacles to improvement are so great.

That this tendency should be particularly prominent in the

younger age groups is natural. It is both to be expected and to be welcomed. The main feature of the youth situation is perhaps the combination of current dependence with the expectation of an early assumption of responsibility. I think that evidence has been presented above that this conflict is accentuated under present conditions. The current youthful indictments of the present state of our society may be interpreted as a kind of campaign position, which prepares the way for the definition of their role when they take over the primary responsibilities, as they inevitably will.

It seems highly probable that the more immediate situation is strongly influenced by the present phase of the society with respect to a certain cyclical pattern that is especially conspicuous in the political sphere. This is the cycle between periods of "activism" in developing and implementing a sense of the urgency of collective goals, and of "consolidation" in the sense of withdrawing from too active commitments and on the whole giving security and "soundness" the primary emphasis. There is little doubt that in this meaning, the most recent phase (the "Eisenhower era") has been one of consolidation, and that we are now involved in the transition to a more activistic phase.

Broadly speaking, youth in a developing society of the American type, in its deepest values and commitments, is likely to be favorable to the activistic side. It is inculcated with the major values of the society, and strongly impressed with the importance of its future responsibilities. At the same time, however, it is frustrated by being deprived of power and influence in the current situation, though it recognizes that such a deprivation is in certain respects essential, if its segregation for purposes of training is to be effective—a segregation which increases with each step in the process of differentiation. A certain impatience, however, is to be expected, and with it a certain discontent with the present situation. Since it is relatively difficult to challenge the basic structure of the youth situation in such respects (e.g., as that one should not be permitted to start the full practice of medicine before graduating from college), this impatience tends to be displaced on the total society as a system, rather than on the younger generation in its specific situation. From this point of view, a generous measure of youthful dissatisfaction with the state of American society may be a sign of the healthy commitment of youth to the activist component of the value system. However good the current society may be from various points of view, *it is not good enough to meet their standards*. It goes almost without saying that a fallibility of empirical judgment in detail is to be expected.

The task of the social scientist, as a scientific observer of society, is to develop the closest possible approach to an objective account of the character and processes of the society. To him, therefore, this problem must appear in a slightly different light: he must try to see it in as broad a historical and comparative perspective as he can, and he must test his judgments as far as possible in terms of available empirical facts and logically precise and coherent theoretical analyses.

Viewed in this way (subject, of course, to the inevitable fallibilities of all cognitive undertakings), American society in a sense appears to be running a scheduled course. We find no cogent evidence of a major change in the essential pattern of its governing values. Nor do we find that—considering the expected strains and complications of such processes as rapid industrialization, the assimilation of many millions of immigrants, and a new order of change in the power structure, the social characteristics, and the balances of its relation to the outside world—American society is not doing reasonably well (as distinguished from outstandingly) in implementing these values. Our society on the whole seems to remain committed to its essential mandate.

The broad features of the situation of American youth seem to accord with this pattern. There are many elements of strain, but on the whole they may be considered normal for this type of society. Furthermore, the patterns of reaction on the part of American youth also seem well within normal limits. Given the American value system we have outlined, it seems fair to conclude that youth cannot help giving a *relative* sanction to the general outline of society as it has come to be institutionalized. On the other hand, it is impossible for youth to be satisfied with the status quo, which must be treated only as a point of departure for the far higher attainments that are not only desirable but also obligatory.

Clearly, American youth is in a ferment. On the whole, this ferment seems to accord relatively well with the sociologist's expectations. It expresses many dissatisfactions with the current state of society, some of which are fully justified, others are of a more dubious validity. Yet the general orientation appears to be, not a basic alienation, but an eagerness to learn, to accept higher orders of responsibility, and to "fit," not in the sense of passive conformity, but in the sense of their readiness to work within the system, rather than in basic opposition to it. The future of American society and the future place of that society in the larger world appear to present in the main a *challenge* to American youth. To cope with that challenge, an intensive psychological preparation is now taking place.

REFERENCES

1. Willard Hurst, *Law and Social Process in United States History* (Ann Arbor: University of Michigan Press, 1960), ch. 2.

2. To sociologists, the frustrating aspects of a favorable situation in this sense may be summed up under the concept of "relative deprivation." See Robert K. Merton and Paul Lazarsfeld, *Continuities in Social Research: Studies in the Scope and Method of The American Soldier.* Chicago: The Free Press of Glencoe, Illinois, 1950.

3. See Talcott Parsons and Robert F. Bales, *Family, Socialization and Interaction Process* (Chicago: The Free Press of Glencoe, Illinois, 1955), especially ch. 1.

4. For example, I am quite certain that the general level of academic achievement on the part of students of Harvard College and the Harvard Graduate School has substantially risen during my personal contact with them (more than thirty years).

5. The emancipation of components that were previously rigidly controlled by ascription is of course a major feature of the general process of differentiation, which could not be detailed here for reasons of space.

6. S. N. Eisenstadt, *From Generation to Generation.* Chicago: The Free Press of Glencoe, Illinois, 1956. See also his paper in this issue of *Dædalus.*

7. For an analysis of this complex in the society, see Winston R. White, *Beyond Conformity.* New York: The Free Press of Glencoe, 1961.

ARTHUR J. GOLDBERG

Technology Sets New Tasks

EACH YEAR, in the month of May, the Secretary of Labor surveys the occupational outlook for our young people and summarizes the main facts in two open letters. One letter is addressed to college graduates who soon will be joining the labor force permanently. The other goes to youth who will be seeking summertime jobs.

Each year the emphasis is increasingly on the importance of educational preparation for employment.

Only yesterday a high school diploma was considered an adequate certificate for a young man or woman applying for his or her first job. Today a college degree is the criterion in virtually all the occupations with a future, and post-graduate achievement is becoming increasingly important.

The prospects for the dropouts from school already are bleak, and rapidly becoming more so.

Our population will increase by an estimated 30 million in the 1960's, but there will be no increase at all in the number of jobs for unskilled workers.

The youth problem involved here was pinpointed by President Kennedy in his May 1962 address to the United Automobile Workers. Observing that in this decade 7 to 8 million of our young people will leave school before they finish, the President said:

One out of every four under the age of twenty today is unemployed. Every analysis looking to the future—and this concerns your sons and daughters—shows that the great need will be in the sixties for those with skills and those with educations. The great lack—the most difficult places to find work in the sixties—will be for those boys and girls without a good education and without training. And we want to make sure that every American has a chance to develop his talents. Education is basic to the preservation of a democracy.

There's the nub of the occupational problem facing members of the Class of '62, and those who are coming after them, in orienting themselves to the industrial community. Change, the basic characteristic of the scientific and technological society into which they were born, presents them with a double challenge. On the one hand, they are under particular pressure to major in public affairs related to the whole occupational complex. On the other hand, they are at the same time required to intensify their concentration on their chosen speciality in business or professional life. They are going to jobs in a time when conditions place a larger premium on action to make sure "that every American has a chance to develop his talents." Clearer understanding of education's profound relationship to the preservation of our democracy has become a new imperative for all of us.

That today's youth must devote more attention to perfecting themselves in a particular speciality may be a clearer imperative than the need of youth to concern themselves with the complicated subject of general employment. However, the tendency of our technological age to generate extensive and sometimes jarring change in the economic environment makes it steadily more difficult for one to maintain an exclusive preoccupation with his individual problems.

Forces long maturing in our society have come to a conjunction which requires a broad expansion and revision of our educational system—a change bringing our training operations more into adjustment with the new industrial tempo. It is a change that must be accomplished in the period now at hand, and its accomplishment imposes larger responsibilities on the individual citizen of our democracy.

We know that the rising rate of job openings for youth who are skilled requires a change in the momentum of dropouts from school, a change in the momentum of large numbers of youths who are unskilled and unlearned. For this to happen, there must be a change toward more education and toward better relationships between schools and industry. And there must be a determination and delineation of what are the basic vocational requirements for young persons who are to enter this era of rapid technological and economic change.

We need, in the first place, the kind of basic vocational education in our schools that will make both the high school and the college graduate *employable*. We need, in the second place, the kind of education that will make them more *flexible*, more readily adaptable to change in the fluid industrial society that we have.

In its full dimensions, the task of education in preparing the skilled population required by our new technology is threefold, and equally urgent on all counts.

1. We must have expanded programs for our colleges and universities to enable them to meet the rising demands of occupations requiring the highest education and training. During this decade the trend toward increasing professional and technical occupations will continue.

2. We must expand and improve our education and training facilities to meet an increasing demand for skilled craftsmen. At the same time, we must have facilities and programs which step up the conversion of unskilled and semi-skilled workers into better trained workers. This is a necessity which grows in urgency as the number of unskilled jobs stays about the same while our population rises. All studies show that the relative number of unskilled jobs will continue to decline over the long term.

3. Improvement in the quality of education for our entire population is patently essential to our national success in the difficult transition period through which we are now passing. We need not only greater capacity and versatility in technical skills but also an increasing measure of knowledge in the humanities for successful management of our maturing industrial democracy. In all of this, education is the touchstone of achievement. It scarcely need be added that an increase in the dollar support for our schools and teachers is indispensable to forward progress.

Today's young men and women have many reasons to ask what is being done and what more can be done to develop our educational system to the level where it can meet all the demands the scientific revolution is placing on it. The question is expressed with particular urgency by their generation.

Looking at the current occupational facts, as set forth in the Department of Labor's *Occupational Outlook Handbook*, we find that the most critical situation exists in the one profession upon which we must ultimately depend for educational improvement. Shortages of qualified teachers exist at all levels of education.

There are also, not surprisingly, unfilled demands for scientists, technicians, engineers, administrators and managers, doctors and nurses, and other specially trained workers, and this discrepancy grows more pronounced as the educational lag continues.

More pressing in its immediate effect is the manifestation of this educational deficiency in unemployment, which is acute among youth. Although young people under twenty-five years of age currently represent 11 per cent of the labor force, they make up 21 per cent of the unemployed. Nearly a million young men and women today are out of school and out of jobs. They represent roughly a fourth of all who are unemployed. Lack of the required skills is a very large factor in the failure of these young people to locate jobs. Dropouts from school lead the procession to idleness.

Of the 26 million young people who will enter the work force in the 1960's, it is estimated that 7.5 million will leave school before they are graduated from high school, and 2.5 million will not advance beyond the eighth grade.

A situation that already is serious will steadily become more aggravated unless we act promptly to upgrade the marketable job skills of a large segment of our youth, as well as to create conditions which will generate more job opportunities.

Better educational preparation for employment will not of itself open doors to jobs for all who need them, of course. We must have expanding production—a higher economic growth rate than we have had in the past—if we are to achieve the goal of full employment. This consideration does not, however, diminish the critical importance of the occupational training factor. A training system which increases the employability of more of our workers will also contribute substantially to greater industrial efficiency and productivity. It is, therefore, pertinent to consider what the federal government is doing and seeking to do as its essential contribution to the total national effort in this field.

The Manpower Development and Training Act of 1962, which is being administered by the Department of Labor's new Office of Manpower, Automation and Training, extends large assistance to states and localities in the training of unemployed workers for available jobs.

Training of unemployed workers is an important feature of the Area Redevelopment Act of 1961.

The Department of Labor's Bureau of Apprenticeship and Training, which this year marks its twenty-fifth anniversary, works with employer and labor organization groups in a unique cooperative effort to update and upgrade the skills of workers. Thousands have improved their proficiency and their earning capacity in these programs.

123

We have had vocational education for years, but not of the kind nor on a scale that would help substantially to meet the problem of the worker's displacement by technological and economic change as it has developed in our time. The job-oriented programs of the Manpower Development and Training Act represent an essential step toward the needed expansion of America's training system.

Federal, state, and local agencies, public and private institutions, and individuals all have essential functions to perform in this cooperative endeavor. Implementation of the Act is, quite literally, projecting the American partnership onto higher ground in the effort to secure equal opportunity for all who are able and willing to work.

These programs have been designed with the purpose of giving our states and localities needed assistance to proceed with programs based on local needs and local initiatives. How much of the task the states and localities will eventually be able to assume entirely on their own, no one can now foretell. What is clear is that our states and localities must have federal assistance—technical and financial assistance of the kind made available in the three-year Manpower Development and Training Act—in order to get the great project started on a national basis.

One part of the Manpower Development Act provides a special program for counseling and training of youths from sixteen through twenty-one years of age. In addition, the Administration is taking a leading part in a nationwide mobilization of state and local agencies and groups, both public and private, for expanded community services to youth. To perform essential liaison and coordinating functions in this phase of the activity, President Kennedy established in 1961 the President's Committee on Youth Employment.

From its headquarters in the Department of Labor, the Committee on Youth Employment works with facts and ideas to widen the public's awareness of youth's problems and to increase public support of undertakings in youth's behalf. Recommendations from the Committee are being drafted by six subcommittees concerned with (1) private and public responsibility for developing job opportunities for youth, (2) preparing in-school youth for occupational employment, (3) labor, management, and education's role in training out-of-school youth, (4) counseling, guidance, and motivation, (5) laws affecting youth employment, and (6) problems of youth in large city slums. The national Committee devotes particular attention to the organization of state and local Youth Committees, who form the leadership corps in the field. They work to provide counseling and job placement services to youth, and to advance training programs and work projects supported by local officials and other community leaders.

Public interest in the Manpower Development Act usually concerns itself first with how many unemployed workers will receive training or retraining under its provisions. It is estimated that there will be 400,000 trainees over the three-year period. We need to realize that, along with improvement in worker skills, this program is significantly enlarging our knowledge of the nation's manpower requirements and resources.

A major feature of the Manpower Act is its provision for a broad program of research studies. In addition to identifying current and prospective manpower shortages in order to determine occupations in which training is needed, these studies are providing detailed analyses of both the benefits and the problems stemming from expanding technology. They seek to establish methods of detecting in advance the potential impact of technological developments on the nation's work force. They are directed to appraising the adequacy of our country's manpower development efforts with a view to recommending needed adjustments. They are looking into the practices which impede or facilitate the mobility of our workers.

Under this Act we have established a National Advisory Committee of ten members (appointed by the Secretary of Labor, and representing labor, management, agriculture, education and training, and the general public). The Committee has the responsibility of making recommendations to the Secretary of Labor to help him carry out his duties under the Act, and of assisting in the organization of committees or groups in plants, communities, regions, and industries to further the purposes of the Act.

In our manpower studies we receive many forceful confirmations of the fact that, as President Kennedy commented at a news conference in February 1962, the major domestic problem of the 1960's will be the achievement of "full employment when automation is replacing man."

One of the principal contributions to fuller understanding of automation's challenge and promise is coming from the President's Advisory Committee on Labor–Management Policy. In its first report to the President, submitted on January 11, 1962, the Committee directed fresh attention to the new frontiers that are opening in this area.

In essence, the report constitutes a three-point statement of principles which our democratic society must follow in order to achieve orderly progress in the age of automation. All of the Committee's twenty-one members (a tripartite group representing labor, management, and the public) agree on these fundamental points:

1. Automation and technological progress are essential to the general welfare, the economic strength and the defense of the Nation.
2. This progress must be achieved without the sacrifice of human values.
3. Achievement of technological progress without sacrifice of human values requires a combination of private and governmental action, consonant with the principles of a free society.

This is an extremely difficult assignment for this or any other generation, and the recognition of that difficulty was manifest in the deliberations of the Committee itself. Though there was Committee unanimity on objectives, some of the Committee members entered separate views on specific means of safeguarding job security while seeking maximum technological progress.

There can be little doubt that in attempting to realize the full promise of automation and solve the dilemma of worker displacement that comes with it, we are wrestling with one of the great contradictions of our history. It is equally clear, however, that we will not meet the central economic test of our time until we have found the answer to technological change "without the sacrifice of human values."

That answer will be found, we may be sure, only through further general enlightenment.

We are at a stage when we must have a policy for education and technical training that fits the whole range of America's needs and aspirations at high tide in the scientific revolution.

Recognition of that fact has shaped the policies of the federal government. Its Manpower Development and Youth Employment programs are part of a whole design for bringing the education and the occupational facility of our citizens up to date in the race with technological change and increasing social complexity.

Central to that design for democratic progress in this era of rapid technological change are the Administration's programs for federal aid for public school construction and teacher training, federal loans for construction of new academic facilities, and federally-financed college scholarships.

We are, unfortunately, still locked in debate over some of the major parts of this program. The debate is concerned largely with how to distribute the federal aid and how much to give. There is little controversy over the need for the expansion in facilities and improvement in teacher training that would be provided under the Administration's proposals. Meanwhile, the national urgency in education grows, as does the Administration's determination to get this job done.

The Administration has set its hand to other important tasks in this area besides the improvement of our schools and colleges and the encouragement of manpower training.

Greater use of the training tool is being made under the Administration's new public welfare program. Enacted by Congress in the 1962 session, this program carries provision for expanded rehabilitative and preventive services to persons who are dependent or who would otherwise be dependent. "This measure," President Kennedy said in signing it into law, "embodies a new approach—stressing services in addition to support, rehabilitation instead of relief, and training for useful work instead of prolonged dependency." It is the most far-reaching revision of our welfare program since its enactment in 1935.

A new program for federal aid to adult education, aimed at banishing adult illiteracy from America, was sent by the Administration to the Congress in 1962. Today we have 8 million men and women aged twenty-five and over who have completed less than five years of schooling. Of this number, more than 2 million have had no schooling. To an important extent, the adult literacy program sought by the Administration would serve as an essential complement to our Manpower Development and Youth Employment programs. Opportunity to learn the three R's is needed by many adult workers who now lack the foundation for the upgrading of skills that would be provided through the training programs of the Manpower Act. As older workers are upgraded, they ease the competition for jobs at the beginner level.

Along with the questions of elementary, secondary, and higher education, adult education, vocational education, and job training programs, the role of the arts in our national life is commanding more attention. The encouragement to the arts and artists that is coming from the White House reflects another profound change at work in America. It is a sign of growing national maturity when our national leadership advances the creative arts for greater consideration along with education and technical training programs designed for a highly industrialized society. These are all parts of the culture which makes a nation great.

In the last two years we have seen many other important evidences of a broadening national concentration on our job problems and our other economic concerns. Among other things, we have been called on to think and act in larger terms with respect to area redevelopment, foreign trade expansion, tax incentives for business investment, tax reform, public welfare, health care for our elderly citizens under Social Security, and public improvements. In most cases, proposals for new federal initiative in these areas have already been translated into action.

We are, in all, moving more purposefully at the federal level to deal with the basic issue of what kind of preparation our whole population must have at this advanced stage of our development as an industrial and urban society.

This paper is, to be sure, largely concerned with federal programs and initiatives. However, the matters discussed here serve to state the dimensions of the problems with which we will be dealing individually and locally—and in each of the fifty states—as well as nationally in this decade.

Regarding the approach to this common task, one thing is very clear. The provision of education is primarily a local responsibility, and a larger assumption of local initiative in providing adequate salaries, staff, and facilities for improved education and training is a prerequisite.

It is also clear that the federal government increasingly faces the challenge to initiate and to lead in national undertakings in this field. With that challenge goes the challenge to fashion programs that will serve to maintain and strengthen the unique federal-state-local partnership which has evolved in this free land for the advancement of our people's great public purposes. A basic test of every measure must be that it meets our traditional democratic rule that the federal contribution should assist, promote, and stimulate a larger exercise of local initiative and local responsibility.

Success in this enterprise requires a growing appreciation in our country of education's enormous contribution to our economic growth, and of education's yet unrealized potentials in that respect.

One-half of the growth in American output in the last fifty years has resulted from factors other than increases in capital and man-hours worked, as is pointed out in the current report of the President's Council of Economic Advisers. Education is one of those factors.

Education's contribution to output is shown by the established fact that income—the measure of individual contribution to production—tends to rise dramatically with educational attainment. The other end of that principle is only too clear: there is little contribution and scant income when educational attainment is, for all practical purposes, nonexistent.

Its beneficent effect on the earning capacity of the individual is, of course, only one part of education's total contribution to our economic progress. This is something we are beginning to assess more accurately as we proceed in the Cold War competition with Communist imperialism, and with efforts to promote order in a world undergoing vast alteration.

Momentous changes in the free world provide more and more evidence that industrial society and democratic society are natural twins, and education is a great unifying and vitalizing force in this union. Their unique compatability has been underscored in recent years by such significant developments as West Germany's stunning economic rebound, postwar Japan's rise as an industrial democracy, Western Europe's sweep to new economic heights and a greater trading union, and the great economic surge that is occurring in all the industrial nations that are independent.

Economic stagnation and decline in Communist East Germany, and scarcities elsewhere behind the Iron Curtain, serve to make the lesson clearer: political institutions which are most suited to the nature and the free spirit of man are also most congenial to scientific progress, technological development, and economic achievement.

In the light of the historical evidence that economic and social growth flourishes best in lands where education and academic freedom are most honored, we need continually to ask ourselves if we are doing enough to create a modern educational system.

Is the education of the citizen preceeding at a rate that accords with the rate of scientific, technological, and economic change?

Are we growing fast enough in awareness that an industrial system so complex and so sophisticated as ours must have workers who are increasingly skilled?

Are we making adequate provisions for the fact that a self-governing political system so intricate and so interdependent as ours must have a citizenry that is increasingly knowledgeable at all levels?

Do we appreciate sufficiently the fact that the most fruitful competition in our system of competitive free enterprise is competition in the realm of ideas?

Are we fully cognizant of the fact that equal opportunity in education is a precondition to the full development of a truly dynamic, creative society?

Are we, finally, entirely aware that a mature industrial democracy demands of its citizens increasing competence in the social and political sciences as well as expanded technological skill and more scientific genius?

To more efficiently administer and develop the democratic system which supports our industrial order, we must have a system of education which increasingly spreads enlightenment on the nature of man and society. We must have a national education policy which stimulates growing cultivation of the qualities of imaginativeness, flexibility, criticism, and receptiveness to new developments. Our greatest tool is the ability to think, and our economic development in the scientific age clearly requires that we place a higher value on humanistic and sociological education as well as on training for scientific, technical, and other occupations.

In all, the challenge of change facing this generation, as it relates to conditions affecting our people's careers and occupations, is a full-size affair. Today's young people are joining the work force at a time when holding a job entails more homework—more application to studies in both the private and public concerns involved in employment—than ever before in our history.

Unlike the youth of certain earlier times, they are, I would assume, under no illusions that they have completed their education when they leave school or college to start their careers. They know, or should know by now, that the pressure and the tension will not let up, or at least that the scientific revolution, the population explosion, and the worldwide struggle for freedom will not slow down for the remainder of this century.

They have grown to man's estate in a period when events should have given them a notably keen insight into the fact that education is the only means by which true revolution and genuine liberation can be obtained. Since this generation started going to school, it has lived through changes which demonstrate the truth of the saying that civilization is a race between catastrophe and education.

They may feel at times that they are being put upon—that history has selected them for a stiffer study course than is their just due. I believe, however, they will take lasting satisfaction from the knowledge that they have the great privilege of being young in the time when scientific man, industrial man, and democratic man are coming of age together.

REUEL DENNEY

American Youth Today
A Bigger Cast, A Wider Screen

*It takes a kind of shabby arrogance to
survive in our time, and a fairly romantic
nature to want to. These are scarce
resources, but more abundant among
adolescents than elsewhere, at least to
begin with.*

---

*Spies moving delicately among the enemy,
The younger sons, the fools.*

GERTRUDE STEIN once explained that the United States is the oldest
country in the world. The United States, she meant, has acquired
the longest experience of modern society. In a *New Yorker* article
of a few years back, Dwight Macdonald told us that the United
States had been the first to develop the concept (not yet accepted in
Europe) of the "teen-ager." His article suggested that the United
States had by now acquired the longest experience with the subculture of youth as it develops in a modern society: our densely
populated youth culture is in some ways the oldest, the most advanced, and the most distinctive in the world. It has also been said,
almost in the same breath, that one reason the young in the United
States do not organize themselves as a social and political movement
is that they are already organized by our society as a body of affluent
consumers.

How much these youthful social forms of ours will become a
model for other societies is still uncertain. Meanwhile, there is agree-

ment that American young people constitute something of a new social type, even while there is disagreement as to what that type is. In the years after World War II, American self-consciousness invented so many new "generations" of the young that the word itself now suggests a period hardly longer than the college life of a college class. Yet it is agreed about the American young that their numbers, homogeneity, prosperity, relatively prolonged dependency, socialization and education (full career comes much later for most, even if marriage comes earlier for many), and their increasing specialization in the labor force constitute their basic and unique generational experiences. Whether they are being prompted by this encounter to substitute new values for old or whether they are simply casting older values in newer forms, is another question.

As a spy in the country of the young, I am sure that my generation's view of its own youth contributes by contrast to any description of what youth is today. We make a contribution simply by recalling the passion with which our mothers selected spinach for our childhood meals, instead of letting us pick our own nutrients, on the basis of TV science, from an early age. We can speak also of the gratitude we felt later on, when our fathers coached us in the grandstand baseball catch in the vacant lot without signing us into the Little League the first time we held tight onto the ball. Still later, our depression days provided us with a modern version of log-cabin origins—we have had to make our way from hardscrabble times in our youth. We even look back to the life of those days as if it were bigger, tougher, more challenging, and more appealing than life is today. If we admit that all these recollections are partly mythical and do not necessarily prove that our survival until now is a mark of merit, they do, as they say, establish our viewpoint. Part of that viewpoint is a conviction that things are not as various as they were.

Isn't it easier to call to mind a typical young person in the United States today than it would have been when I was between twelve and nineteen, from 1925 to 1930? There were not nearly so many of us then, nor did we bulk so large as a proportion of the total population. Yet through no fault of our own we were more diverse. Differences in family income were greater than they are today, and they were reinforced by distinct class, ethnic, and generational styles of life. Today the bilingual and the foreign-accent family, for example (except among the Latins), are becoming pressed flowers in the American album of nostalgias. The very speech of the American young, with its centripetal drift toward the mid-Western speech pattern that has become the American standard, tells what has happened. Within the great internal market of the nation, the automobile, the electronic wave, and a variety of other forces have served an equalitarian tropism by making young people richer than they have ever been, and rather more like each other than they have been since the Civil War. That date marks the last previous era when American young people, with the exception of Negroes and Indians, constituted a homogeneous population: North European and English-speaking farm-bred folk with lower grammar-school literacy and a predominantly Protestant world view.

For the more than 61 million people under eighteen to constitute such a large percentage of the total population under sixty-five as they do today is also a cyclical return to an aspect of earlier historical experience. At the time of the signing of the Declaration of Independence, half the nation was under eighteen. We do not touch this proportion today. But the infant and first-year mortality rate has fallen from one out of three in 1880 to a few in a thousand. By 1965, when the great 1947 peak in the postwar birth rate must be felt even more than it is now, it will have the effect of doubling within a decade the number of people annually arriving at the age of eighteen—from two million in 1956 to four million in 1965. Within the next few years, about eighty percent of this group will be in high school. About one-third of these will leave high school without graduating, the boys to enter the army or the labor market, the girls to enter the labor market or to marry without entering it. Of the more than one and one-half million who graduate from high school, about half will go on to register at college, more young men than young women. It can be expected that the roughly twenty-five percent of the sixteen to twenty-four age group who now are registered in college will increase. The age of extended socialization is in full swing.

One of the odd effects of extended socialization is that it seems to us older ones to be connected as much with a downward extension of age-graded roles and a general widening of privileges as it does with the postponement of a full career as an adult. In my day, it was chiefly the farm youth who learned as a matter of course to drive a car early in life, and not many of the young were car-owners. Today, driving and car-ownership are fully accepted and more or less unexciting stages in the rites of passage of almost all young men and women. The high-school student of today has taken up the ceremonial burden of graduating in a manner befitting the college graduate of earlier years. He expects to be equipped with the cars, white jackets, carnations, orders for corsages, and all the other equipment of role and status that in my day as a stripling were reserved by the Florist's Association, the Cadillac ads, the pages of *Vanity Fair,* and the apparel shops, for well-off adults and a small number of country club sprouts. In 1925, the earliest pairs of trousers were provided for sons of the family as they were readying for college or a job. Today, the earliest pairs of trousers are tailored for infant girls who do not even know enough syllables to protest the absence of a working fly. All these precocious privileges appear at the lower end of the age scale, even while the upper end is displaying the novelty of the father of three who is being worked through college by his father-in-law and through his graduate thesis by his wife. The downward shift of age privileges may be a narrowly middle-class occurrence. It is conservatively resisted by some upper-income groups. In some lower-income groups on the move, the shift works in the other direction. But the largesse of male privilege to both sexes seems general and as strong in South Harlem as in Scottsdale, Arizona.

Perhaps it is more a matter of the purse than anything else. The twelve-year-old suburban girl whose shopping list read "water-pistol, brassiere, and permanent" was demonstrating not merely the tremors of transition but also purchasing power. If you can buy as an adult, you *are* an adult. And one of the main things about adult buying in the United States is that you do not need ready money to do it. "Grow up now and pay later" might have been the motto of the youth who quickly spent a fantasy four thousand on a credit-card spree. *Life's* researchers have tallied and budgeted the slacks, records, and tennis rackets owned by young American men and the slacks, records, and tennis dresses owned by young American women. The ten-billion-dollar stake of the young purchaser in the total consumer expenditures of the nation has been examined and certified by Eugene Gilbert. With the exception of the children of some three million families in the United States that survive on a brutally low income, there are few young men who cannot scrape up enough to buy a second-hand car.

It is really not so long ago that most young people in industrial society were sacrificed to the ideal of capital formation. The transformation of the youth from a family asset as labor to a family liability as student-consumer was generalized earlier in the United

States than anywhere and has by now had far-reaching effects. In the United States, an industrial-minded middle class with an ethos of postponed gratification took to the idea of prolonged education for everyone, and the notion of a public investment in the young was generally accepted without much grumbling by the urban working classes. Unions were glad to see young workers standing in school lunch lines rather than crossing picket lines. Much of the public economy of modern youth was "socialized" by the benefits of the American school system, even while capitalists were proclaiming that God had given the country to the men of property to run. Moreover, the cult of youth in the United States may be attributable partly to the speed with which this nation was able to build the skills of men into machines. That process made each older generation of skilled workers feel more vulnerable to obsolescence than their grandfathers could have felt, and, by contrast, dramatized the plastic readiness of youth for the encounter with new ways of production and consumption. Youth as an industrial institution has been developing since the mid-eighteenth century; now the whole bureaucracy of modern industry and war pivots to an increasing degree on the rationalized training, recruitment, and placement of young people who, because of the nature of modern economic development, will not reach full productivity until the maturation of sectors that are being planned ten and twenty years in advance. The division of labor is felt to be reaching almost into the kindergarten.

For the youth as student and as worker in the very near future, two trends that would once have had to be regarded as contradictory will be felt at the same time. Because of our postwar growth in population, the number of persons under twenty-four in the work force will far exceed any previous figures for this group. On the other hand, because of our enlarged educational goals and ambitions, more young people than ever before will be held until a later age out of the labor market. The educational preparation of the young poses three issues of great importance that are troubling us now: inequality of opportunity, premature specialization, and the glorification of the average. The young boy or girl in the United States today is offered an education whose high-average achievements are marred by segregation, compromised by the belief that all vocational talents can and should be identified at an early age, and clouded by the number of high schools in which the best is too close to the average. Segregation is present in the North as well as the South, of course. Northern segregation is based on class rather than caste and is not openly defended by many. These are cold comforts to progressive school districts whose standards can be lowered by ill-prepared Negro students; they are even colder comforts to Negroes lined up to get something better than substandard treatment by the public purse.

As for specialization, it is notable that American high schools offend less in this respect than do many colleges. The American high school student can find basic disciplines and teachers of great ability in many places today. Yet more public high schools might be even better than they are, resembling New Trier, in Winnetka, Illinois, for example, if their parent constituencies were more informed, and if their core curriculum were more loyally classical. There is an increasing belief among some middle-class parents that a first-rate secondary school may be even more important than a good college.

American higher education presents a widely mixed scene in which vocationalism still runs strong. More students take degrees in business than in either the sciences or the humanities, and the average college graduate in the United States is culturally illiterate. More than half of them, for example, read less than seven books a year. Graduate and professional schools are of good but varied quality, and it is natural that they should tend to seek narrowly professional competences in their graduates. They show disloyalty to the general educational scene when they insist (as they often do) that their preconceptions and prerequisites should divert students from a broad education in the earlier school years.

In the face of the educational bureaucracy, with its persistently laggard standards in both private and public schools, the better college student of today often feels in lock step. He knows that most American higher education is too industrially organized around classrooms, grades, and hastily contrived examinations. Moreover, the American habit of interference in the private lives of students, induced partly by American coeducation, partly by public education, partly by the theory of *in loco parentis,* is undignified for those who run universities and unedifying for students. It has now been complicated by shocking governmental pressures on universities to supply information about students which they have received in professional confidence. The comparative dependency of American college youth is a scandal in the Western world: in California recently, young university students were indifferently treated as infants and as swine by police assigned to keep order at hearings of the House Un-American Activities Committee. It is perhaps for partial escape from these conditions that many a student of today turns toward a college culture of work-study-sociability. This sector, if one takes into account the large number of students who are involved, the scale of their efforts, the variety of their part-time occupations, and the linkage of their work with the newer leisure in the United States, amounts to a new thing on the American scene.

Even the recent recession did not prevent about two million students from finding summer work and thus continuing that dramatic re-entrance of youth into the labor force that occurred during and after World War II, reversing the youthful joblessness of the Depression. The new employment of the American student on a part-time basis involves every industry, skill, and role, but much of its impetus comes from the seasonal, the service, and the tourist trades. The young American student in high school or college bases his expenditures on a budget in which public scholarship, private scholarship, family support, and self-earnings are often all present. This youthful work economy is not, by and large, vocationally directed, even though it may serve as a way of trying out possible occupations. It is rather a form of paid sociability combined with study, an existence in which the student-waiter brings some of the campus to the resort and uses the pool after hours. Even if it does not reach these pleasant heights, it possesses a general, national quality that is to be recognized anywhere. It is dependent upon the newer service-centered economy that we associate with the resort, the marina, the bowling alley, the camp, the summer campus, the summer music school, the baby-sitting guild, the campus research project, the young straw-hat theater, the youth hostel, and all the rest.

These new social scenes of experience-seeking youth have been created by public education, the increase of the average family income, the rise of leisure expenditures, the urban public programs of recreation, the national parks, the extension works of the farm bureaus and the state universities, the expansion of audiences for art and music, the increasing "brow" tone of resorts, the rise of the conference trade (every topic has its conference, just as every trade has its convention), and a variety of other forces that seem to have dropped a Chautauqua tent over the country from Eastport to San Diego. It is no chore for the nation to support thousands of young ski-bums, some from mansions and some from shanty towns. The right to self-supporting adventure at work, taken away from youth by automation and unionization at the beginning of the century, has been given back with a premium in the latter part of the century. Thousands are still excluded from this subsidized playground by reason of race or other conditions of servitude, but the remarkable thing is how many are in. Even without the glamor, there are more spots in the United States today where kids can work in a beanery in the morning and study Sanskrit or folk dancing in the afternoon than there were saloons in the Old West or sweat shops in the Old East Side.

These scenes of youthful work-play-and-study are pervasive but not universal, and even the young who enjoy them most are prone to anxiety. The reasons vary. One of the main reasons seems to be that, although wealthy and conservative nations such as the United States want their children to be responsible and obedient rather than aggressive and full of initiative, they hesitate to admit it. The extraordinary interest of the American parent in the early occupational specialization of the young male is one sign of this. With the decline of status associated with class and family the vacuum has been filled by status associated with the professional badge. Again, in *The Lonely Crowd,* David Riesman, Nathan Glazer, and I observed that many young people seemed to care more about "fitting in" with their friends and their society than ever before. A study of this theme among both students and their parents, by John and Mathilda Riley, seems to show that parents are hardly less interested in the conformity of their young than the young themselves. What lies under these appearances of complaisance?

In search of the self concept, some of the young prefer to be known, after seventeen or so, by the somewhat starchy title of "young adults." On the other hand, a fluent loose-jointedness, restlessness alternating with catatonic repose, can be found in all classes of the young. "Like *this* or like that, *man,*" so popular a construction in their speech, suggests a feeling for life that is both tentative and metaphorical. Youth can be so different from itself, as it moves from role to role, that it still maintains, even in a society devoted to publicity, a great capacity for concealment. This masquerade, and the range of youth, and the pace of change, make it difficult to generalize about the young. There was the young high-school graduate who recently dared the disapproval of his community by rejecting the award of a veterans' group whose attitudes he did not honor. On the other hand, there are the hot-rodders who perpetually hum "Transfusion," by Nervous Norbus:

> Toolin' down the highway doin' 79
> I'm a twin-pipe poppa
> And I'm feelin' fine.

Hey, dig that!
Was that a red stop sign?
(Sound of crash)
Transfusion, Transfusion!

It seems that young people are living through a change in the
character structure of our society and that they show it in their response to public events. The modern mass media have made it possible for a glamorous academic to use cribbed answers on a TV show.
Seven-eighths of the students polled at his own school (Columbia
University) sympathized with the fallen hero, and many found occasions to make excuses for him. An almost equal number of mid-
Western high-school students responded in more or less the same
way. Many of the grounds offered by the students for minimizing
the star's offense seem lax and complacent. On the other hand, there
is something worthy of attention in the observation by some of these
students that the difference between entertaining fiction and intellectual fraud in the mass media is an ambiguous matter. The difference depends in part on conventions linking the program with the
audience and on the sophistication of the audience itself. It seemed
fantastic to many of these young people that the audience could
take the quiz programs as anything but a circus in which the phoney
was an admitted part of the game.

A comment by Edgar Z. Friedenberg in *The Vanishing Adolescent* seems to me to be very much to the point here. "It must be
taken for granted that in many respects our conception of integrity is
obsolete; we include in it many ways of feeling and acting that acquired their social significance under social conditions that no longer
exist." He goes on to say that "individualism, which led to success in
a society dominated by the economic necessities of industrialization
and empire, is a poor model for the young today." I myself am particularly impressed by the difficulties in judgement induced by the
unequal socialization of the older and the younger people in the
mass media. The older people are the more literal, more censorship-
minded, and more out of touch in their response to the mass media.
They often label as cynical a youthful response to the mass media
that is based on skills they do not possess. This occurs in each generation, of course, but there is reason to think that the very rapid increase in the younger educated groups has recently widened the gap
between the fifteen-year-old and the fifty-year-old in our society. The
second most popular public visiting place for service men in New
York, after the Empire State Building, is the Museum of Modern Art.

139

This suggests that young people in general are increasingly the participants in a cosmopolitan culture from which most older members of the society, because of provincialism and lack of training, are excluded. An apt illustration of this is the fact that young people have generally had ample opportunity to enjoy the *double entendre* of a new jazz or calypso record by the time adult ears have understood it and banned it from the air.

Admittedly, there is much evidence that many of the youthful population are notable chiefly for their traditionalism. Readers of reports by H. H. Remmers and D. H. Radler of Purdue University have been shocked by their findings in recent years. They show that a general, low evaluation of tolerance for diverse viewpoints in the adult population (as shown by Samuel A. Stouffer in *Communism, Conformity and Civil Liberties*) is also quite widely distributed among the high-school students of the country. These young people are "traditional" in the sense that they continue a habit of disrespect for traditional American liberties of conscience, free speech, and the press. Even if it is not quite clear whether they always understood what they were saying in their answers to Remmers and Radler, the number of them who took an authoritarian position is disturbing. There are so many of them for each one of the Southern white girls who followed their beliefs about the sit-ins right into the sheriff's office. Indeed, it is quite possible that a hell-fearing intensity in personal morality can cool off, not toward a greater liberality in view of the self and of others, but toward a defensive self-righteousness. This apparently defensive form of illiberality, as found among the youth of some rural areas, is perhaps associated with the strain between facts and appearances in the political life of the regions. The co-authors of *Small Town in Mass Society* suggest that the less these towns control their own political destiny, the more they talk about the virtues of local self-rule. In such situations, the young are presented with adult models who preach nineteenth-century individualism while reaching for twentieth-century subsidies.

Edgar Z. Friedenberg, considering this matter, makes one of the major contributions, along with Erik H. Erikson and Paul Goodman, to our understanding of this age period. He says that adolescence in the older sense no longer exists except as a remnant. In our own youth it was still possible to think of adolescence as the period in which the young person formed his identity by a meaningful and dynamic conflict with parents and the older generation. This is what adolescence actually consisted of, in social-psychological terms. In terms of social commandments, this is what society was satisfied that it should be. The older and the younger, the richer and the poorer, the male and the female, were all more or less agreed on this. This view of the age period depended, however, on the general acceptance of a clear-cut idea of adult identity and maturity. Today, when this concept no longer exists so clearly (so autonomous an adult would be deviant in our culture) the growing-up process that assumed it falls to the ground also. It is the very absence of the drama of confrontation with the adult that troubles many modern adolescents, more than its presence would.

These comments may help us to understand why so many collective portraits of young people sometimes sound so oddly dated. They often take it for granted that a well-defined conflict between youth and age is being acted out. This, in turn, seems related to the assumption that the study of differences of "opinion" between younger and older groups in the society can be taken at face value. I doubt that the discovery of youth can proceed in terms of a questionnaire vocabulary alone: "Do you believe in personal immortality? What is wrong with parents?" What makes the inquiry even more difficult is that what youth is and becomes is determined more by the small group of the creative young than by its majority members. When read properly, a poem, a painting, or a piece of scientific work by a young person can tell us more about his generation than an article entitled, "What Youth Wants" can. In most schools, we must also remember, a respectable percentage of the students are brighter than their teachers, and what the keenest twelve-year-olds already know about pure science is inaccessible to more than three-quarters of all the rest of us. A young actor such as James Dean could say more about postwar youth than most young people themselves or their parents and teachers. It is of some significance that one of the scripts through which he spoke, *Rebel Without A Cause*, dealt with a middle-class, father-son relationship that was unsatisfactory because it was marked by slackness rather than moral tension.

These considerations of the weakening super-ego do not, of course, exhaust the question of the models that young people find or do not find for themselves. Neither do they suggest all the social consequences of the models they choose. Among the delinquents of our period, the search for the adult to be emulated is clearly a desperate and bitter quest. In the first place, as New York and Chicago well know, the "far-out" delinquent is statistically associated with unacculturated families living in the crowded apartments in which middle-class life as we know it is impossible. Second, as Sheldon and Eleanor Glueck suggest, a delinquent is generally the product of a family life that deviates from the model "good family environment"—that of the ambitious and responsible mother married to the encouraging and permissive father. Third, the American youth takes to delinquency in many cases because of the wide gap between culturally approved goals and the culturally provided means for his attaining them.

Recent research in Chicago indicates that the way out of delinquency is mediated by the nature of the neighborhood. In the "law-abiding" neighborhood, the prospects are for reform. In the neighborhood that provides the middle management men of organized lawlessness, the prospects are that the delinquent will be taken into the syndicate system or barred from lawlessness entirely. In the poor neighborhoods, that are victimized by organized lawlessness and corruption, the prospects are grimmer. Those who refuse to reform slide toward increasingly intense conflicts with society. Of all the youthful varieties of deviancy, this is the most automatically self-destructive and the hardest to deal with. There is no overwhelming evidence that the large federal programs now contemplated as a counterattack on youthful crime can succeed while the families and the school experiences of young offenders continue as now.

The question of the general character structure of the young, in contrast to previous generations, is complicated, of course, by changes in the sexes' roles. The convergence in these roles makes it possible to say that much of what has been suggested in this article so far is as true for young women as for young men. The intersexual, like the interethnic, feelings of the young used to play across a social and sociable space rather crudely organized in terms of class and ethnic endogamy, economic values, and an array of class-spaced and sex-separated church-goings, sports, and other occasions for entertainment. Today the sociable activities in which the sexes are separated still continue their long decline. Such things as the men's club are retained more firmly by the few Greek-Americans of the city corner than they are by the rich or the educated.

The slow but steady increase in privileges for women (including the privilege of working hard both at home and on the job) is a mixed matter in the United States. For one thing, it derives partly from a frontier scarcity of women which no longer prevails. For another, it is marked by considerable inequalities for women of various social and ethnic groups. What is even more important, the newer freedoms of women, including young women, has been vitiated by the absence of ideal types. The adulation of Mrs. Roosevelt among widely varying social groups might be considered evidence for the hunger felt by American women (and men too) for what an aristocratic tradition might have bequeathed to American women had it ever existed in strength. The most general development for young women is that they are increasingly permitted to enter the labor force at higher and higher levels, but that few have the resources, especially economic ones, to control both a job and a home. Despite the movement toward younger marriages, a declining divorce rate, and the newer "cooperative" marriage (in which the family life gets rationalized like a business, or parceled out in roles like an *atelier*) many young women seem to regard caring for husbands and trotting after children as the unexciting payments they make on an insurance policy. The tendency toward younger marriages also seems to have threatened the postschool period of work-and-leisure that women's magazines define as every American girl's birthright as a single person.

There are some signs of relief in the younger feminine generation at the widened range of interests of American middle-class males. Nothing seems to have failed so grandly on the American scene of the last twenty years as the frontier, rural, hair-on-the-chest version of masculinity. This is surely related to the declining demand for labor in the extractive and farming enterprises—we can produce with less than ten percent of our work force enough food to make young Americans, like their parents, generally overweight. If the muscular pose still persists on the more or less rural and Western co-ed school scene, it is on the defensive even there. Young men and women are aware that the ambitious young fellow of today is as likely to be aiming for a position on the charity or ballet board as he is for a golf club membership—the more so the higher his corporate ambitions are.

They are also aware that scientists consider themselves licensed to feel bored with other scientists who do not play the classical guitar or translate verse. The culture of the small town, the range states, and the small business enclave still fights for the old male ideal by dint of bars furnished with longhorns and fast talk about oil wells. Although some of this is a genuine remnant of frontier culture and not a sagebrush stage set for the farm lobby, it is remarkable how much it needs to be sustained by mythology. No young males enjoy the imagery of the ruggedly individualistic more than those of the arid rural states do—states that would hardly break even were it not for federal subsidies of farm products, metal and water resources, and the returns of tourist bars, dude ranches, gambling tables, and graft on the public lands. One of the most observable shifts of fashion among young people since World War II is that toward the subtle, the Italianate, the pale, and the pensive. The wholesome frontier blonde and the stubbled football player have gone out of fashion as hero and heroine.

As a drama, the life of American youth today presents itself in a bigger cast than ever, playing on a wider screen. Youth plays a role to which the word "juvenile" can no longer be applied, because, while this word once meant someone up to fourteen or so, it now means a grammar-school youngster—or a book written for one. There is evidence as to young people in the decline of books intended for the twelve- to sixteen-year-old—today *Tom Swift* would not pass muster with much younger readers, and *Ann of Green Gables* could not compete with *Mademoiselle*. We might ask then, in a more general way, whether new interpretations by young people of perennial hero figures are a sign of coordinate changes in their self-image.

Not long ago there appeared a new, young, New York production of *Hamlet*. This version of *Hamlet* treated what used to be interpreted in the character of Hamlet as a lack of action and an incapacity for action, as a kind of action in itself. We may be on our way toward a new *Hamlet*, one that will overturn the nineteenth-century conviction that Hamlet's trouble was simply too much thought, and the early twentieth-century conviction that his trouble was simply too much Oedipus complex. Perhaps young people need only look at the heroes of Samuel Beckett to redefine Hamlet as being furiously activist, even extrovert! Could this new *Hamlet* be evidence that youth projects on this hero the semblance of its own bureaucratized self? Youth lives in a world in which physical action and labor have been replaced by brain work and in which clear-cut personal goals are less evident than the subdivided, bureaucratic adjustments to role. Perhaps this is a world in which any second thought, capable of paralyzing any action, is still preferable to principled decisions that carry overtones of fanaticism.

By and large, it is only groups that have been released from the struggle for physical and social survival that possess the time for

brooding. The decrease in challenge, as my colleague David Riesman phrases it, is a significant part of the experience of contemporary young people. The increasingly extended period of education
and socialization has its constrictions, but they are made of cotton
wool. Many of the most recent interesting elements in the lives of
young Americans seem to be centered on the matter of the balance,
or lack of balance, between the advantages and disadvantages of the
long socialization period that our economy not only makes possible
for all but virtually obligatory for most.

One response to the absence of directly recognizable challenge
is an attitude of indifference toward the conventionally structured
roles and functions offered by the society. One young student I know
has called this mood "institutionalized indecision." It was melo-
dramatized for a while by the Beats. "Me for the pads where you
don't have to choose" might have been their variant on a famous line
by Robert Frost. Although some of the gurus of the Beats had not
been young since a Joan Crawford movie called *Flaming Youth,*
there actually were some young ones in the circuit and their aim was
part-time work for a part-time life with a part-time ego. In Venice,
California, the new Bohemian seems at times to prefer the job that
constitutes at once the shallowest commitment of one's self and the
broadest satire on the rat race. (Some of them, for example, report
their jobs as those of deodorant testers.) This retreat from the job
market seems to have been dramatized partly in clothes: the sandals
and bare feet would scare a plant-safety man out of his head. Even
certain aspects of clothing that were originally tinged by industrial
utility (for example, the minimally tailored trousers favored first by
the cowboy and later by industries wary of loose clothing) have become so attenuated that they suggest the hose that once went with
the breeches in the times of Henry VIII. Surely no young man totter-
ing about with a Henry VIII profile, with short coat hung over
shoulders as if it were a flaring cape and legs as thin and irresolute
as green noodles, can be expected to lift a tool.

Extended dependency and extended education, along with the
concomitant bureaucratization of life that accompanies these, are the
most significant factors in the life of young people in the United
States today. My own students, however, are the first to remind me
how many of the young do not finish high school. Observing the joblessness of the young in the recent recession, they called my attention
to the strained relationship between the young blue-collar entrant to
the labor market and the unions. The recession multiplied the im-

pact of seniority, not only on employment but also on union membership. There are signs in Detroit of a coolness in the young toward the unions that exclude them. Even though this puts a pressure on some of the unemployed young to use their time to upgrade themselves in the occupational system by acquiring more education, it does not make higher education dominant in the lives of most of them. The big institution in most of these young blue-collar lives, my students tell me, is the Job.

My students are both right and wrong. They are right that the Job looms as large as anything in the mind of the boy who leaves high school so that he can support a car. They are wrong, I think, in underestimating the speed with which unskilled jobs are vanishing on the American scene and with which public higher education is generating collegiate values and styles of life among all the young. When I was in high school, in the late 1920's, the college boy and the college girl were already heroes of the movies and the popular imagery. Yet even then it was apparent that the movie college of the 1920's was actually a high school landscaped as a country club and inhabited by high-school mentalities who plucked their eyebrows and shaved. Now, almost all young people share some of the style of life that older magazine editors still tend to identify narrowly with the American campus. For, of all the things that are affecting the life of youth today, one of the most influential is the way in which the educational enterprise has acquired economic scale and significance. Education has become not merely a "defense" stock in the sense of being a region's fiscal defense against recessions in other industries, but perhaps even a pump-primer, like a highway plan.

A paradox of social circumstances is that new solutions inaugurate new dilemmas. There are proportionately more young Americans in our population pyramid today than there have been since the early nineteenth century. Their dependence on the fully working population is both longer and more expensive than ever before in human history. On the other hand, these very young people are increasingly committed to making later contributions to the support of a larger and larger demographic attic of older people. Again, although the affluence of the young continues to rise, the satisfactions of the young well-to-do have not been entirely assured by their comforts, and the purchasing power of the poorest young people has not kept pace with the expectations incited by the levels of expenditure that they see every day around them. Moreover, while it is true that today's young are committed to an increasingly longer experience with school, the educational institutions themselves, especially at the higher levels, have become departments of modern adult life itself, rather than anterooms to it.

The same pattern of change occurring on several levels at once is provided by the marriage customs of youth. They marry somewhat earlier than their parents did, and as a result they are "on their own" in life at a stage when they have less experience. Yet some of the confidence that enters into the earlier marriage is associated with the confidence of the young in our economic institutions. These institutions in turn are increasingly dependent on extraordinary scales and schedules of public expenditure, installment-plan living, and an economy of the mixed private-public variety in which the free play of competitive forces is much limited by welfare policies and monopolistic privileges. In this society, young people are "on their own" only in a sense that many of the young people of the 1890's would have found quite unrecognizable. Most of today's young marrieds seem to be aware of this historic development. Their acceptance of it seems at times almost complacent to my generation.

The occupational life that supports the young family is also being affected, in bewildering ways, by changes in the occupational structure. On the one hand, there is the emphasis on the security that is generally associated with ambitions for specialization. With fewer unskilled workers, everyone seems to be a specialist of one sort or another, if only by grace of a nomenclature that now gives the title of "Junior Researcher" to the "clerk" of 1920. On the other hand, the nature of modern industry is such that the number of jobs for which more than a few weeks' training is actually needed increases fairly slowly. Industrial practices, professional restrictions, educational prerequisites often seem to have much more to do with the elaboration of specialized job functions than do the requirements of the job themselves. It is felt by many parents and their children that movement into the more educated and professionalized occupations is an unalloyed gain of income as well as of status. Fewer seem to be aware, as George Stigler points out, that the rise of skilled employees leads to a natural reduction of their comparative returns, unless the demand is as dynamic as the supply. Since the demand for the higher trades and services is heavily structured by political and cultural policy, the prospects are that their development may well be subject to wide fluctuation in the future.

It is only recently in human history that we have deployed machinery in order to take most people out of personal service to others. As a result, the low status associated with many of these occupations depresses their recruitment and development. The market of the young actor is undercut by governmental policies in amusement taxes at the same time that the market for the young farmer is coddled by agricultural subsidies. The trades of the tailor and the cook and the gardener, and even the teacher, are devaluated by the culture at the very time in the development of the culture when the actual market for these distinctive services might be expanding. In short, the old rigidities in the labor market and the old frictions in labor mobility seem to have been replaced by new ones. The confusing speed with which occupational ambitions change is exemplified by the inability of the advertising business to recruit new personnel among the urban and the better-educated young people of today. The public cross-purposes with respect to the rise of newer trades, markets, and ways of life are demonstrated by the right-hand, left-hand contradictions of legality in New York City. The New York fire statutes, which are insufficient to contend with the overpopulation of slum apartments, are sufficient to drive young painters out of the cheap lofts which are the functional prerequisites of their work and their lives.

Comparable cross-currents are to be found in the way in which young people are prepared for the life of partial leisure imposed on them by modern industry. A good many of the activities of the young, including their high-school and college courses, are preparations for leisure. They range from instruction in how to run an outboard motorboat to courses in Shakespeare. Some of the former are better, in the sense of delivering the goods, than the latter. Yet the suspicion remains that these developments are more often involved with the expansion of markets (textbooks and gasoline) than they are with development and discipline of the human needs that are the soul of the markets. The phonograph disc has two faces: one side is for the disc jockey, who is paid to make markets (whether he makes them or not) by the industry. Only the other side, and not all that, is left to represent the part played by the disc in elevating and differentiating the American taste for music and the spoken word.

It is in connection with the issue of freedom versus limits, and in connection with the activist mentality versus the expressive personality, that important problems of understanding lie. The poor Negro youth may be taken as the example of one who lives under the pressure of limits overimposed. His difficulty is compounded by the fact that, whereas he was formerly permitted to react to this situation by an emphasis on expressiveness rather than on activism, he is decreasingly allowed to take this stance, either by his own people or by the whole society. He is forced to give up some aspects of proletarian values at the same time that he is prevented from fully taking up the middle-class values that would confirm this revolution in himself.

The privileged white youth of the suburbs may be taken as an example of one who lives in a certain vacuum of freedoms. Here, the traditional content of American activism has little meaning, because the economic problem is not pressing, and the school is only a bit more so. From such groups come many of the young, especially the moody males, who react against any ideal of activism as such. For them, the only way is down to the old expressiveness of the past proletariat (the "white Negro") or up to the newer expressiveness of the sciences, the arts, the crafts, the performing arts, the design trades, and perhaps even areas that used to be considered as feminine enclaves or monopolies. Their identity is more and more formed, not by a conflict with their immediately older generation, but by an encounter with the forms of imagination and articulation bequeathed by the aristocratic traditions of the past. For example, this is the type of the young American actor, who institutionalizes some lack of super-ego formation by transferring the formation to later years in life and to the process of being led by a director. In its most aggressive form, this is the type of young American who strenuously revives the older American "Song of Myself" versus the bureaucrats, while at the same time, as Beat, denies its traditional industrial content and shifts his ambitions toward the expressive trades in the American division of labor. Here, the search for the self-imposed limit is connected with an attempt to achieve a transference to the cultural values of the tradition rather than those of the fathers. Perhaps this is the reason for the beards.

Most of America's young people lie between these two types. Many of them, neither frightened by dreadful freedoms nor oppressed by social tyrannies, seem aware of the tenor of our time. They seem to understand that it is not the increased power of the young in modern society but the decreased power of the older to manage it that is the core of the matter. Even this middle group, therefore, gives greater signs than our own youth did of laying the older optimism and activism aside. In certain ways, modern American young people seem to walk on eggs more than any generation in the twentieth century. Their talent for the "delayed reflex" may prove to be one of our main resources in the coming culture and politics of the nuclear age.

Acknowledgements: The writer has received valuable comments and suggestions from James Short (in lectures at the University of Chicago); Esther Moir DeWaal, of Cambridge University; Marc Swarz, of the University of Chicago; Stephen Karpf and Nancy Taylor, students at the University of Chicago.

The first quotation at the head of this article is from Edgar Z. Friedenberg's *The Vanishing Adolescent* (see the Bibliography below). The second is reprinted by permission from James Agee's dedicatory poem to Walker Evans in James Agee and Walker Evans, *Let Us Now Praise Famous Men* (Boston: Houghton Mifflin Company, 1960), Book 2, p. [5].

BIBLIOGRAPHY

1. Berelson, Bernard, *Graduate Education in the United States.* New York: McGraw-Hill Company, 1960.

2. Conant, James Bryant, *The American High School Today.* New York: McGraw-Hill Company, 1960.

3. Erikson, Erik H., *Childhood and Society.* New York: W. W. Norton and Company, 1950.

4. Friedenberg, Edgar Z., *The Vanishing Adolescent.* Boston: Beacon Press, 1959.

5. Gilbert, Eugene, "Why Today's Teen-Agers Seem So Different," *Harper's,* November 1959, pp. 76-79.

6. Ginzberg, Eli (editor), *The Nation's Children.* New York: Columbia University Press, 1960.

7. Glueck, Sheldon and Eleanor, *Unravelling Juvenile Delinquency.* New York: The Commonwealth Fund, 1950.

8. ——— *Delinquents in the Making.* New York: Harper & Brothers, 1952.

9. ——— *Predicting Delinquency and Crime.* Cambridge: Harvard University Press, 1959.

10. Goodman, Paul, *Growing Up Absurd.* New York: Random House, 1960.

11. Henry, Jules, Jack Lyle and Gregory Stone, "Notes on the Alchemy of Mass Misrepresentation," *Studies in Public Communication,* No. 3, 1961.

12. Lang, Gladys Engel and Kurt, "Van Doren as Victim," *Studies in Public Communication,* No. 3, 1961.

13. Lieberman, Myron, *The Future of Public Education.* Chicago: University of Chicago Press, 1960.

14. Lipset, Seymour M. and Leo Lowenthal (editors), *Culture and Character: An Evaluation of the Work of David Riesman.* Chicago: The Free Press of Glencoe, Illinois, 1961.

15. Lipton, Lawrence, *The Holy Barbarians.* New York: Julian Messner, 1959.

16. Macdonald, Dwight, "Profile," *The New Yorker,* 22 November 1957.

17. Mailer, Norman, *Advertisements for Myself.* New York: G. P. Putnam's Sons, 1959.

18. Mayer, Martin, *The Schools.* New York: Harper & Brothers, 1961.

19. "New Ten-Billion Dollar Power," *Life,* 31 August 1959, pp. 78-85.

20. *New York Times,* 6 June 1960, p. 43.

21. Nervous Norbus, "Transfusion," quoted in Robert H. Boyle, "Car Cultists," *Sports Illustrated,* 24 April 1961, p. 72.

22. Plunkett, Margaret, "Geographic Mobility of Young Workers," *Occupational Quarterly,* (United States Department of Labor), 1960.

23. Remmers, H. H. and D. H. Radler, *The American Teen-Ager.* Indianapolis-New York: Bobbs Merrill Company, 1957.

24. Riesman, David, "The Jacob Report," *American Sociological Review,* 1958, *23:* 732-738.

25. —————— "Found Generation," *American Scholar,* 1956, *25:* 421-436.

26. —————— Nathan Glazer and Reuel Denney, *The Lonely Crowd.* New Haven: Yale University Press, 1961.

27. Stigler, George, *Trends in Employment in the Service Industries.* Princeton: Princeton University Press, 1956.

28. Stouffer, Samuel A., *Communism, Conformity and Civil Liberties.* Garden City: Doubleday and Company, 1955.

29. Sussman, Marvin, "Activity Patterns of Post-Parental Couples," *Marriage and Family Living,* 1955, *17:* 338-341.

30. —————— "Help Pattern in the Middle-Class Family," *American Sociological Review,* 1953, *18:* 22-28.

31. Vidich, Arthur, and Joseph Bensman, *Small Town in Mass Society.* New York: Doubleday and Company, Anchor Book No. A216, 1959.

JOSEPH F. KAUFFMAN

Youth and the Peace Corps

It is not thy duty to complete the work
but neither art thou free to desist from it.

SAYINGS OF THE FATHERS

As THIS IS WRITTEN, the Peace Corps is twenty months old. A brief
fifteen months have elapsed since the first volunteers were sent overseas. Nevertheless, over four thousand are now at work in thirty-
eight countries of Asia, Africa, and Latin America, and by the fall
of 1963 we can expect this number to be increased to eight thousand.

The Peace Corps was created by Presidential Order on March 1,
1961. The United States Congress enacted legislation in September
1961 and voted funds to finance its operations. Robert Sargent
Shriver has directed the agency from its inception.

The purposes of the Peace Corps, as stated in the Peace Corps
Act are

to promote world peace and friendship through a Peace Corps, which
shall make available to interested countries men and women of the
United States qualified for service abroad and willing to serve, under
conditions of hardship if necessary, to help the peoples of such countries
and areas in meeting their needs for trained manpower, and to help promote a better understanding of the American people on the part of the
people served, and a better understanding of other peoples on the part
of the American people.

It is clear from the above that the goals of the agency lie not
only in performing urgently needed tasks but in increasing friendship and understanding. The opportunity to "do a job," linked with
the essential humanitarian goals of the Peace Corps, has captured
the imagination and idealism of countless citizens.

Any American citizen over eighteen was declared eligible to
apply for Peace Corps service. No upper age limit was set. Persons
with dependents under age eighteen were ruled ineligible. In the
case of married couples, both husband and wife were required to

serve together so that families would not be separated. Each volunteer would receive intensive training in the language and culture of the country to which he was going as well as training in the skills required for his job.

It was with these purposes and principles that the Peace Corps began its task of achieving its bold promise.

It is far too early for one to assess the Peace Corps, either in order to make claims of its full success or to extrapolate from its brief experience knowledge which can be confidently applied to the subject of youth in general. Nevertheless, if the Peace Corps had within it the seeds of failure, it would have failed by now. Its experience both here and abroad, however limited, may well have considerable relevance to the subject of youth in our day.

The initial response of the American people to the idea of the Peace Corps is of primary significance. We are not yet free of the stereotyped thinking about our youth, and there were cries of consternation in some circles about the idea of young men and women, naive and flabby, interfering in the foreign relations of our nation. Too many politicians, and social scientists as well, accepted all too quickly the notion that American young people did not have the zeal, idealism, dedication, or toughness to meet the challenge the Peace Corps idea represents. They were mistaken.

In a recent debate in the United States Senate concerning the expansion of the Peace Corps, Senator John Sparkman of Alabama commented:

I believe the Peace Corps has tapped an asset we have always had but never used, except in time of war. I speak of the drive and dedication of the young men and women of this nation. Time after time, during military hostilities, we have drawn on our youth and they have served and died for our country.

Now in time of relative peace, we have realized that the same qualities that have made our fighting men the world's best, also make them, and their companions of the opposite sex, the best exportable evidence of what this nation is.

They have offered their talents, their skills and two years of their lifetimes for the performance of national service abroad in order that they might play a personal and individual role in the promotion of world peace and friendship.

Thousands of persons have volunteered for Peace Corps service abroad. Each is a unique individual and it is difficult to categorize them in any manner other than an obvious and general one. Nevertheless, it may be of interest to get a fuller picture of those who have been selected to serve in this manner.

Peace Corps volunteers have come from every state in our nation and their ages have ranged from eighteen to sixty-nine. The vast majority, however, range in age from twenty-one to twenty-five. Men outnumber women by two to one, although this figure is distorted somewhat by the fact that there are fewer jobs overseas for women than for men; this balance may be altered in the future. In terms of education, almost all Peace Corps volunteers have pursued their education beyond the high school level and some 75 to 80 per cent are graduates of colleges, possessing at least a bachelor's degree.

Volunteers are serving abroad in a variety of professional, technical, and occupational jobs. More than half are serving as teachers in elementary, secondary, and teacher-training institutions. Besides teaching, the major tasks are in the fields of agriculture, community development, nursing, and public health. Nevertheless, the service rendered and the role played by the volunteer in the host country cannot be limited by the usual description of these tasks. By its very nature, the Peace Corps places the volunteer in a living situation which necessitates his becoming involved in the community in which he performs his work. Therefore, many activities remain undefined and are created by the volunteer within his own situation.

The Peace Corps volunteer is a unique American abroad. His presence has been requested but he is not an "expert." He works under the supervision of host-country nationals, lives on the level of his co-workers, under their laws and without special privileges or immunities. He is not a spokesman for the American government but a free individual who is voluntarily serving in areas where few Americans have ever lived or worked before. It is his own work, behavior, sensitivity, and self-responsibility that reflect credit or discredit on the free society which has produced him. Almost without exception, the performance to date has been outstanding. The best evidence of this is that each of the fifteen countries in which Peace Corps volunteers first served has asked for additional volunteers.

Sympathetic critics often asked whether Peace Corps service isn't really a form of "escape"—a postponement of the assumption of normal responsibility, of the acceptance of reality. I think it is not; but the answer is not that simple.

"Escape" need not always have a negative connotation. The single persons in their twenties whom I have been privileged to see in Peace Corps training programs and at work in Africa are not running from anything, although they are most assuredly responding to something. Perhaps overseas service for two years does fulfill a need for them, but our society has failed to provide alternative, socially approved, useful ways in which this need could be met.

In fact, the very success of the Peace Corps—the response it has evoked from the best of our youth—may well be an indication that something is lacking in our society as it relates to youth. If there is a flaw in society's perception of what our young people are thinking, feeling, or dreaming, then I believe this flaw has been perceived and used by the Peace Corps idea. If other meaningful channels had been open to these young people, the Peace Corps might not have worked.

In earlier times, the college years served as a kind of "moratorium" on the acceptance of responsibility; there was time, relaxation, reflection, and study unrelated to vocational goals. Today, from the early high school years on, every move of work, study, or pleasure remains within the context of building a record (curricular and extracurricular) directly related to future goals of college, graduate or professional school, fellowships, job opportunities, and the like. In fact, expending one's time, energies, and abilities in ways unrelated to future personal goals is, in many circles, considered irresponsible.

In many ways today's college experience encourages and perpetuates a kind of egocentrism which is not always a natural frame of reference for the college age group. Instead of being assisted in relating one's life to the community, to humanity, or to a whole, one is encouraged to work for oneself.

No one can deny the importance of individual responsibility, but an emphasis on working solely for one's own success, with all its ramifications, is obviously not wholly satisfactory as a milieu for creative and positive identification or the drive to give to and assist the whole.

I do not mean to imply that Peace Corps experience will not be used to enhance future careers—quite the contrary; but for most it is not another means to a functional end. It is seen as a value in and of itself, a broadening life experience that requires no defense and deserves no ascription of guilt. If it seems to some to be an interlude outside of the race for personal career or economic advancement, to many others it is a means of personal fulfillment. Such means are all too scarce today and we need more, not fewer, of them.

It is my own personal observation that Peace Corps service offers several remarkable opportunities to which young men and women are eager to respond. I shall attempt to list several of these and to illustrate them by quoting from communications received from volunteers in the field.

1. The volunteer senses or perceives that he will have an opportunity to apply what is *theoretically* the ideal in our value system but which rarely has an opportunity to be expressed—namely, the blending of sensitivity, patience, and the humane and ethical virtues with the qualities of courage or manliness. He perceives that he will have to be both sensitive and tough, prudent and adventurous, intelligent and practical, strong and yet aware of his own limitations.

2. The volunteer has the opportunity to feel honestly needed and to perform vital tasks that require his skill and training. One can *give*, in terms of both energy and good will, without any fear that one is in a juvenile game, as is all too often the case when youth participates in many adult activities in both college and community here. The need to know that what one is doing is important, and *makes a difference*, can be met for most volunteers by their service abroad.

3. The opportunity to learn, about others as well as oneself, is clearly a part of the Peace Corps experience and young people have responded knowingly to this fact. I speak here not merely of learning about another people, another culture, another language, although this is significant and important in its own right. I refer particularly to the opportunity the Peace Corps presents to youth to test themselves, to discover their own intellectual capacities, and those capacities of character and personality which demand testing and definition.

To help illustrate these points, let me cite the thoughts of some individual volunteers as expressed in their letters to us. The first is from a young man serving in the Peace Corps in Chile. From a lengthy letter I excerpt the following:

> I am anxious to return to Rio Negro because I enjoyed December there. My life was hectic but not unhealthy. In one period of nine days I slept in eight different places—once sprawled on the floor of a chapel in my sleeping bag. I traveled from place to place on foot, horseback, bus, truck and sometimes in the jeep of the North American "Padres" in Rio Negro. I ate well enough and although a great portion of our Peace Corps contingent has suffered from intestinal disorders, I remain one of the lucky ones.
>
> Another big reason for wanting to return is that I like the Capesinos and have grown to respect them a great deal. Their endurance for hard work, their ability to live off the land without any of the things which we

call "necessities," their love of music and the generosity which prompts
them to slaughter one of their few sheep for the meal they are serving
me, leaves me filled with admiration.

A young female college graduate serving in Pakistan expresses
in a letter to a friend back home the kind of sensitivity which we
know exists in our youth but which rarely has an opportunity to
express itself.

My job is a little overwhelming at this point. I have been working on
a syllabus for the anatomy and physiology course I'll be teaching. Now
that it is ready, I have to start on the others. I have also had a tour of
the city health clinics and the leper colony. After my trip to the latter,
all of us here are planning to adopt the leper colony as our pet project.
There are 44 lepers, men, women and children, all living together in
one old Hindi temple. It is situated about ten miles out of town in a
barren, God-forsaken strip of unproductive land. There are no houses for
miles and the lepers seem to be a forgotten people. The nuns and mission-
aries go out regularly with food and bring a little change of scene into
their lives. They have a few chickens and a plot of land to farm but it
still seems too little and a rather sad and futile existence. It will be the
one project we will all be working on together in our free time and everyone is quite enthusiastic about it. Even the fact that we will have to
bicycle the ten miles hasn't seemed to dim spirits any. There is so much
to do! Where do you start! I feel so small and insignificant.
. . . You can't help but become involved but you can somehow find
solace in the little things you do and bend with the rest. Some of the
Volunteers are finding it very hard [to avoid] becoming depressed with
the enormity of it all. None of us say how we really feel. We gripe and
complain, laugh and poke fun, and yet each of us has another side sacred
only to himself. Why are we embarrassed to express how we really feel?
For fear we might sound corny and possibly be laughed at? We are such
a cold culture outwardly and yet we are the softest on the inside. How
we should hate ever being found out!

Another young woman teaching school in a remote area of Ni-
geria indicates that there are numerous rewards which one tends to
easily forget or overlook because it is difficult to assess them. One
observation she makes, however, is particularly telling.

One can almost feel oneself growing and expanding one's prespective
and understanding. In some ways, it is like going through the adolescent
experience again with both its rewards and frustrations. The important
things to come with are three main attitudes: of learning, of unfailing
optimism, and of flexibility.

Finally, there is the letter from a young man teaching school in a
newly independent West African country desirous of more Peace
Corps volunteer teachers than can be provided at this time. This

157

volunteer is teaching mathematics and science to ill-prepared boys on the secondary school level. It has not been easy either professionally or in terms of his living conditions and relative isolation. Yet, after six months on the job he is moved to write the following in response to a request for information and guidance to other volunteers who are training to go to the same country in the near future.

. . . If I seem to be painting a rather dark and gloomy picture of this scene, it is intended, for these are the things for which people should be prepared when they come here. . . . For what else should a Volunteer be prepared? He should be prepared for a delightful, warm, friendly, appreciative and fun-loving people, for a student who is, in the main, industrious and eager to learn. He should also be prepared for the consternation afforded by ill-prepared students, for the nerve-racking frustrations which arise out of incomprehension and consistent failure. He should be prepared for a rewarding experience which will live with him as long as he is on this earth.

Whether any or all of these things will bear any great interest or significance I do not know. I do know that I have come upon a situation which has caused me to stop and completely re-evaluate myself, my ideas about education, and about a person's integral relationship with his culture. I have always held (somewhat idealistically and academically, to be sure) that there is a certain "oneness" about humanity which no amount of epidermal coloring or cultural uniqueness could hide. The last five months in West Africa have done nothing to alter that view, except insofar as it has strengthened it.

As the Peace Corps idea becomes more real and functional I begin to more and more realize what a rare and wonderful opportunity this whole thing becomes for everyone involved. The need for education here is desperate, and the appreciation we have been getting from people on all levels is no less than astounding. Some are a little hesitant to believe that we would give up the luxuries of America, the good jobs, the money and the conveniences for that which West Africa has to offer, but they are nonetheless glad to have us. One student said to me, "I really don't understand why you would want to do this, but welcome." This is not to say that many do not grasp the underlying idealism involved but many find it difficult to believe that it springs from American soil. I think that the best explanation we can give is an operational one—visible and tangible evidence of Americans living and working here for someone else's interest. The question "Why did you join the Peace Corps?" which has plagued us from the beginning, becomes increasingly less and less academic. One no longer has to resort to abstract philosophical arguments and platitudes (true as they might be), for I now find myself in the midst of the answer, surrounded by a situation which cries out in self-explanation. The poverty, the illiteracy, the sub-standard educational opportunities which are rampant in this as well as many other countries are reasons enough for anyone to want to extend his hand and heart in order that these blights might be at least partially effaced.

It now appears likely that as many as 10,000 Peace Corps volunteers will be serving abroad within a relatively short time. As the Peace Corps grows, there will be a steady flow— first in the hundreds, then in the thousands—of volunteers who have completed their two years of service abroad and are home to take up their responsibilities as adults in our school, religious, civic, and business life.

The resources for enriching our educational institutions, as well as industry, government, and labor, are boundless. Will we find the means of tapping this combination of knowledge, experience, and altruism? Perhaps more important, will the sense of commitment and new understanding of returning volunteers touch the nation and infuse it with a reinforced sense of commitment and understanding?

Recently, Professor Arnold Toynbee wrote an article in the *Cyprus Mail* concerning a visit he made to the Peace Corps Field Training Center in Puerto Rico. His closing remarks are very much to this point:

In the Peace Corps the American citizen is deliberately placing himself in the position of the immigrant. He recognizes that, if he is to establish a human relationship with the people of the country where he is going to work, he will have to do this in their language, not in his own. This is a revolution indeed, and it is to be hoped that its effects will spread from Americans who have served in the Peace Corps to other Americans serving abroad and to their fellow countrymen at home.

Here we have put our finger on an aspect of the Peace Corps that is of great national importance for the United States. If the Peace Corps makes a success of its job abroad, it will also serve as a leaven at home. It may revolutionize the American people's attitude towards the non-Western majority of mankind. If it succeeds in doing this, it will have made history at home as well as abroad. It will have given history a turn for the better, and it will have done this in the nick of time.

My purpose here has not been to evaluate the Peace Corps nor to engage in Pollyanna-ish statements designed to recruit new volunteers. I have tried to show simply that the response evoked by the Peace Corps idea has been real and sincere. This response must be taken seriously, for I believe that it contradicts many of our notions and demands our attention.

If we are to encourage and utilize the vast resources of energy and idealism that seek expression, our leadership must be of a quality and character to call forth the necessary response. I am convinced that this is possible in our colleges and universities and in our civic and religious institutions. There are sufficient challenges to which

159

youth can respond—challenges involving thought as well as action. The means and the goals cannot be spurious. The leadership must be of the highest integrity. But given these ingredients, American youth will respond.

Our nation's challenge then rests not only with its youth but with leadership which youth can respect. We have made a good start with the Peace Corps. The private sector affords even greater opportunities for those who care.

KENNETH KENISTON

Social Change and Youth in America

EVERY SOCIETY TENDS to ignore its most troublesome characteristics.[1] Usually these remain unfathomed precisely because they are taken for granted, because life would be inconceivable without these traits. And most often they are taken for granted because their recognition would be painful to those concerned or disruptive to the society. Active awareness would at times involve confronting an embarrassing gap between social creed and social fact; at other times, the society chooses to ignore those of its qualities which subject its citizens to the greatest psychological strain. Such pluralistic ignorance is usually guaranteed and disguised by a kind of rhetoric of pseudo-awareness, which, by appearing to talk about the characteristic and even to praise it, prevents real understanding far more effectively than could an easily broken conspiracy of silence.

Such is often the case with discussions of social change in America. From hundreds of platforms on Commencement Day, young men and women are told that they go out into a rapidly changing world, that they live amidst unprecedented new opportunities, that they must continue the innovations which have made and will continue to produce an ever-improving society in an ever-improving world. Not only is social change here portrayed as inevitable and good, but, the acoustics of the audience being what it is, no one really hears, and all leave with the illusory conviction that they have understood something about their society. But it occurs to none of the graduating class that their deepest anxieties and most confused moments might be a consequence of this "rapidly changing world."

More academic discussions of social change often fail similarly to clarify its meaning in our society. Most scholarly discussions of innovation concentrate either on the primitive world or on some relatively small segment of modern society. No conference is com-

161

plete without panels and papers on "New Trends in X," "Recent Developments in Y," and "The New American Z." But commentators on American society are usually so preoccupied with specific changes —in markets, population patterns, styles of life—that they rarely if ever consider the over-all impact of the very fact that our entire society is in flux. And however important it may be to understand these specific changes in society, their chief importance for the individual is in that they are merely part of the broader picture of social change in all areas.

Even when we do reflect on the meaning of change in our own society, we are usually led to minimize its effects by the myth that familiarity breeds disappearance—that is, by the belief that because as individuals and as a society we have made an accommodation to social change, its effects have therefore vanished. It is of course true that the vast majority of Americans have made a kind of adaptation to social change. Most would feel lost without the technological innovations with which industrial managers and advertising men annually supply us: late-model cars, TV sets, refrigerators, women's fashions, and home furnishings. And, more important, we have made a kind of peace with far more profound nontechnological changes; new conceptions of the family, of sex roles, of work and play cease to shock or even to surprise us. But such an adaptation, even when it involves the expectation of and the need for continuing innovation, does not mean that change has ceased to affect us. It would be as true to say that because the American Indian has found in defeat, resentment, and apathy an adaptation of the social changes which destroyed his tribal life, he has ceased to be affected by these changes. Indeed, the acceptance and anticipation of social change by most Americans is itself one of the best indications of how profoundly it has altered our outlooks.

Thus, though barraged with discussions of "our rapidly changing world" and "recent developments," we too easily can remain incognizant of the enormous significance, and in many ways the historical uniqueness, of social change in our society. Rapid changes in all aspects of life mean that little can be counted on to endure from generation to generation, that all technologies, all institutions, and all values are open to revision and obsolescence. Continual innovation as we experience it in this country profoundly affects our conceptions of ourselves, our visions of the future, the quality of our attachment to the present, and the myths we construct of the past. It constitutes one of the deepest sources of strain in American life,[2] and many characteristically "American" outlooks, values, and institutions can be interpreted as attempts to adapt to the stress of continual change.

Social Change in America

Many of the outlooks and values of American youth can be seen as responses to the social changes which confront this generation.[3] But merely to point out that society is changing and that youth must cope with the strains thus created is to state a truth so universal as to be almost tautological. Social change is the rule in history: past ages which at first glance appear to have been static usually turn out on closer study to have been merely those in which conflicting pressures for change were temporarily canceled out. Indeed, the very concept of a static society is usually a mistake of the short-sighted, a hypothetical construct which facilitates the analysis of change, or a myth created by those who dislike innovation.[4] All new generations must accommodate themselves to social change; indeed, one of youth's historic roles has been to provide the enthusiasm—if not the leadership—for still further changes.

And even if we add the qualifier "rapid" to "social change," there is still little distinctive about the problems of American youth. For though most historical changes have been slow and have involved little marked generational discontinuity, in our own century at least most of the world is in the midst of rapid, massive, and often disruptive changes, and these may create even greater problems for the youth of underdeveloped countries than they do for Americans. Thus, to understand the responses of American youth to the problems of social change, we must first characterize, however tentatively and impressionistically, the most striking features of social change in this country.

Social change in America is by no means *sui generis;* in particular, it has much in common with the process of innovation in other industrialized countries. In all industrially advanced nations, the primary motor of social change is technological innovation: changes in nontechnological areas of society usually follow the needs and effects of technological and scientific advances. But though our own country is not unique in the role technology plays, it is distinguished by the intensity of and the relative absence of restraint on technological change. Probably more than any other society, we revere technological innovation, we seldom seek to limit its effects on other areas of society, and we have developed complex institutions to assure its persistence and acceleration. And, most important, because of the almost unchallenged role of technology in our society, our attitudes toward it spread into other areas of life, coloring our views on change in whatever area it occurs. This country closely approximates the ideal type of unrestrained and undirected technological change which pervades all areas of life; and in so far as other nations wish to or are in fact becoming more like us, the adaptations of American youth may augur similar trends elsewhere.

Our almost unqualified acceptance of technological innovation is historically unusual. To be sure, given a broad definition of technology, most major social and cultural changes have been accompanied, if not produced, by technological advances. The control of fire, the domestication of animals, the development of irrigation, the discovery of the compass—each innovation has been followed by profound changes in the constitution of society. But until recently technological innovation has been largely accidental and usually bitterly resisted by the order it threatened to supplant. Indeed, if there has been any one historical attitude toward change, it has been to deplore it. Most cultures have assumed that change was for the worse; most individuals have felt that the old ways were the best ways. There is a certain wisdom behind this assumption, for it is indeed true that technological change and its inevitable social and psychological accompaniments produce strains, conflicts, and imbalances among societies as among individuals. Were it not for our own and other modern societies, we might ascribe to human nature and social organization a deep conservatism which dictates that changes shall be made only when absolutely necessary and after a last-ditch stand by what is being replaced.

But in our own society in particular, this attitude no longer holds. We value scientific innovation and technological change almost without conscious reservation.[5] Even when scientific discoveries make possible the total destruction of the world, we do not seriously question the value of such discoveries. Those rare voices who may ask whether a new bomb, a new tail fin, a new shampoo, or a new superhighway might not be better left unproduced are almost invariably suppressed before the overwhelming conviction that "you can't stop the clock." And these attitudes extend beyond science and technology, affecting our opinions of every kind of change—as indeed they must if unwillingness to bear the nontechnological side effects of technological innovation is not to impede the latter. Whether in

164

social institutions, in ideology, or even in individual character, change is more often than not considered self-justifying. Our words of highest praise stress transformation—dynamic, expanding, new, modern, recent, growing, current, youthful, and so on. And our words of condemnation equally deplore the static and unchanging—old-fashioned, outmoded, antiquated, obsolete, stagnating, stand-still. We desire change not only when we have clear evidence that the status quo is inadequate, but often regardless of whether what we have from the past still serves us. The assumption that the new will be better than the old goes very deep in our culture; and even when we explicitly reject such notions as that of Progress, we often retain the implicit assumption that change *per se* is desirable.

Given this assumption that change is good, it is inevitable that institutions should have developed which would guarantee change and seek to accelerate it. Here as in other areas, technology leads the way. Probably the most potent innovating institution in our society is pure science, which provides an ever-increasing repertoire of techniques for altering the environment. An even greater investment of time and money goes into applied science and technology, into converting abstract scientific principles into concrete innovations relevant to our industrialized society. The elevation of technological innovation into a profession, research and development, is the high point of institutionalized technological change in this country and probably in the world. And along with the institutionalized change increasingly goes planned obsolescence, to assure that even if the motivation to discard the outmoded should flag, the consumer will have no choice but to buy the newest and latest, since the old will have ceased to function.

But the most drastic strains occur only at the peripheries of purely technological innovation, because of changes in other social institutions which follow in the wake of new commodities and technologies. Consider the effects of the automobile, which has changed patterns of work and residence, transformed the countryside with turnpikes and freeways, all but destroyed public transportation, been instrumental in producing urban blight and the flight to the suburbs, and even changed techniques of courtship in America. Further examples could be adduced, but the point is clear: unrestrained technological change guarantees the continual transformation of other sectors of society to accommodate the effects and requirements of technology. And here, too, our society abounds with planning groups, special legislative committees, citizens' movements, research organizations and community workers and consultants of every variety whose chief task is, as it were, to clean up after technologically induced changes, though rarely if ever to plan or coordinate major social innovations in the first place. Thus, citizens' committees usually worry more about how to relocate the families dispossessed by new roadways than about whether new roads are a definite social asset. But by mitigating some of the more acute stresses indirectly created by technological change, such organizations add to social stability.

One of the principal consequences of our high regard for change and of the institutionalization of innovation is that we have virtually assured not only that change will continue, but that its pace will accelerate. Since scientific knowledge is growing at a logarithmic rate, each decade sees still more, and more revolutionary, scientific discoveries made available to industry for translation into new commodities and techniques of production.[6] And while social change undoubtedly lags behind technological change, the pace of social innovation has also increased. An American born at the turn of the century has witnessed in his lifetime social transformations unequaled in any other comparable period in history: the introduction of electricity, radio, television, the automobile, the airplane, atomic bombs and power, rocketry, the automation of industry in the technological area, and equally unprecedented changes in society and ideology: new conceptions of the family, of the relations between the sexes, of work, residence, leisure, of the role of government, of the place of America in world affairs. We correctly characterize the rate of change in terms of self-stimulating chain reactions—the "exploding" metropolis, the "upward spiral" of living standards, the "rocketing" demands for goods and services. And unlike drastic social changes in the past (which have usually resulted from pestilence, war, military conquest, or contact with a superior culture), these have taken place "in the natural course of events." In our society at present, "the natural course of events" is precisely that the rate of change should continue to accelerate up to the as-yet-unreached limits of human and institutional adaptability.

The effects of this kind of valued, institutionalized, and accelerating social change are augmented in American society by two factors. The first is the relative absence of traditional institutions or values opposed to change. In most other industrialized nations, the impact of technology on the society at large has been limited by pre-existing social forces—aristocratic interests, class cleavages, or religious values—opposed to unrestrained technological change. Or, as in

166

the case of Japan, technological changes were introduced by semi-feudal groups determined to preserve their hegemony in the new power structure. Technologically induced changes have thus often been curbed or stopped when they conflicted with older institutions and values, or these pretechnological forces have continued to exist side by side with technological changes. The result has been some mitigation of the effects of technological innovation, a greater channeling of these changes into pre-existing institutions, and the persistence within the society of enclaves relatively unaffected by the values of a technological era.[7] But America has few such antitechnological forces. Lacking a feudal past, our values were from the first those most congenial to technology—a strong emphasis on getting things done, on practicality, on efficiency, on hard work, on rewards for achievement, not birth, and on treating all men according to the same universal rules.

A second factor which increases the effect of technological change is our unusual unwillingness to control, limit, or guide directions of industrial and social change—an unwillingness related to the absence of institutions opposing innovation. Most rapid changes in the world today involve far more central planning or foreknowledge of goal than we are willing to allow in America. At one extreme are countries like China and Russia, which attempt the total planning of all technological, industrial, and social change. While unplanned changes inevitably occur, central planning means that the major directions of change are outlined in advance and that unplanned changes can frequently be redirected according to central objectives. Furthermore, most underdeveloped nations are aiming at developing a highly technological society; in so far as they succeed, the direction of their changes is given by the model they seek to emulate. Given three abstract types of change—planned, imitative, and unguided—our own society most closely approximates the unguided type. We do little to limit the effects of change in one area of life on other aspects of society, and prefer to let social transformations occur in what we consider a "free" or "natural" way, that is, to be determined by technological innovations. As a result, we virtually guarantee our inability to anticipate or predict the future directions of social change. The Russian knows at least that his society is committed to increasing production and expansion; the Nigerian knows that his nation aims at increasing Westernization; but partly by our refusal to guide the course of our society, we have no way of knowing where we are headed.

The Phenomenology of Unrestrained Technological Change

Man's individual life has always been uncertain: no man could ever predict the precise events which would befall him and his children. In many ways we have decreased existential uncertainty in our society by reducing the possibilities of premature death and diminishing the hazards of natural disaster. But at the same time, a society changing in the way ours is greatly increases the unpredictability and uncertainty of the life situation shared by all the members of any generation. In almost every other time and place, a man could be reasonably certain that essentially the same technologies, social institutions, outlooks on life, and types of people would surround his children in their maturity as surrounded him in his. Today, we can no longer expect this. Instead, our chief certainty about the life situation of our descendants is that it will be drastically and unpredictably different from our own.

Few Americans consciously reflect on the significance of social change; as I have argued earlier, the rhetoric with which we conventionally discuss our changing society usually conceals a recognition of how deeply the pace, the pervasiveness, and the lack of over-all direction of change in our society affect our outlooks. But nonetheless, the very fact of living amidst this kind of social transformation produces a characteristic point of view about the past and future, a new emphasis on the present, and above all an altered relationship between the generations which we can call the phenomenology of unrestrained technological change.[8]

The major components of this world view follow from the characteristics of change in this country. First, the past grows increasingly distant from the present. The differences between the America of 1950 and that of 1960 are greater than those between 1900 and 1910; because of the accelerating rate of innovation, more things change, and more rapidly, in each successive decade. Social changes that once would have taken a century now occur in less than a generation. As a result, the past grows progressively more different from the present in fact, and seems more remote and irrelevant psychologically. Second, the future, too, grows more remote and uncertain. Because the future directions of social change are virtually unpredictable, today's young men and women are growing into a world that is more unknowable than that confronted by any previous generation. The kind of society today's students will confront as mature adults is almost impossible for them or anyone else to anticipate. Third, the present assumes a new significance as the one time in which the environment is relevant, immediate, and knowable. The past's solution to life's problems are not necessarily relevant to the here-and-now, and no one can know whether what is decided today will remain valid in tomorrow's world; hence, the present assumes an autonomy unknown in more static societies. Finally, and perhaps of greatest psychological importance, the relations between the generations are weakened as the rate of social innovation increases. The wisdom and skills of fathers can no longer be transmitted to sons with any assurance that they will be appropriate for them; truth must as often be created by children as learned from parents.

This mentality by no means characterizes all Americans to the same degree. The impact of social change is always very uneven, affecting some social strata more than others, and influencing some age groups more than others. The groups most affected are usually in elite or vanguard positions: those in roles of intellectual leadership usually initiate innovations and make the first psychological adaptations to them, integrating novelty with older values and institutions and providing in their persons models which exemplify techniques of adaptation to the new social order. Similarly, social change subjects different age groups to differing amounts of stress. Those least affected are those most outside the society, the very young and the very old; most affected are youths in the process of making a lifelong commitment to the future. The young, who have outlived the social definitions of childhood and are not yet fully located in the world of adult commitments and roles, are most immediately torn between the pulls of the past and the future. Reared by elders who were formed in a previous version of the society, and anticipating a life in a still different society, they must somehow choose between competing versions of past and future. Thus, it is youth that must chiefly cope with the strains of social change, and among youth, it is "elite" youth who feel these most acutely.

Accordingly, in the following comments on the outlooks of American youth, I will emphasize those views which seem most directly related to the world view created by unrestrained change,[9] and will base my statements primarily on my observations over the past decade of a number of able students in an "elite" college. While these young men are undoubtedly more articulate and reflective than most of their contemporaries, I suspect they voice attitudes common to many of their age mates.

Outlooks of Elite Youth

One of the most outstanding (and to many members of the older generation, most puzzling) characteristics of young people today is their apparent *lack of deep commitments to adult values and roles.* An increasing number of young people—students, teenagers, juvenile delinquents, and beats—are alienated from their parents' conceptions of adulthood, disaffected from the main streams of traditional public life, and disaffiliated from many of the historical institutions of our society. This alienation is of course one of the cardinal tenets of the Beat Generation; but it more subtly characterizes a great many other young people, even those who appear at first glance to be chiefly concerned with getting ahead and making a place for themselves. A surprising number of these young men and women, despite their efforts to get good scholarships and good grades so that they can get into a good medical school and have a good practice, nonetheless view the world they are entering with a deep mistrust. Paul Goodman aptly describes their view of society as "an apparently closed room with a rat race going on in the middle."[10] Whether they call it a rat race or not is immaterial (though many do): a surprising number of apparently ambitious young people see it as that. The adult world into which they are headed is seen as a cold, mechanical, abstract, specialized, and emotionally meaningless place in which one simply goes through the motions, but without conviction that the motions are worthy, humane, dignified, relevant, or exciting. Thus, for many young people, it is essential to stay "cool"; and "coolness" involves detachment, lack of commitment, never being enthusiastic or going overboard about anything.

This is a bleak picture, and it must be partially qualified. For few young people are deliberately cynical or calculating; rather, many feel forced into detachment and premature cynicism because society seems to offer them so little that is relevant, stable, and meaningful. They wish there were values, goals, or institutions to which they could be genuinely committed; they continue to search for them; and, given something like the Peace Corps, which promises challenge and a genuine expression of idealism, an extraordinary number of young people are prepared to drop everything to join. But when society as a whole appears to offer them few challenging or exciting opportunities—few of what Erikson would call objects of "fidelity"—"playing it cool" seems to many the only way to avoid damaging commitment to false life styles or goals.

To many older people, this attitude seems to smack of ingratitude and irresponsibility. In an earlier age, most men would have been grateful for the opportunities offered these contemporary young. Enormous possibilities are open to students with a college education, and yet many have little enthusiasm for these opportunities. If they are enthusiastic at all, it is about their steady girl friend, about their role in the college drama society, about writing poetry, or about a weekend with their buddies. Yet, at the same time, the members of this apparently irresponsible generation are surprisingly sane, realistic, and level-headed. They may not be given to vast enthusiasms, but neither are they given to fanaticism. They have a great, even an excessive, awareness of the complexities of the world around them; they are well-read and well-informed; they are kind and decent and moderate in their personal relations.

Part of the contrast between the apparent maturity and the alienation of the young is understandable in terms of the phenomenology of unrestrained change. For the sanity of young people today is partly manifest in their awareness that their world is very different from that of their parents. They know that rash commitments may prove outmoded tomorrow; they know that most viewpoints are rapidly shifting; they therefore find it difficult to locate a fixed position on which to stand. Furthermore, many young men and women sense that their parents are poor models for the kinds of lives they themselves will lead in their mature years, that is, poor exemplars for what they should and should not be. Or perhaps it would be more accurate to say, not that their parents are poor models (for a poor model is still a model of what not to be), but that parents are increasingly irrelevant as models for their children. Many young people are at a real loss as to what they should seek to become: no valid models exist for the as-yet-to-be-imagined world in which they will live. Not surprisingly, their very sanity and realism sometimes leads them to be disaffected from the values of their elders.

Another salient fact about young people today is their relative *lack of rebelliousness* against their parents or their parents' generation. Given their unwillingness to make commitments to the "adult world" in general, their lack of rebellion seems surprising, for we are accustomed to think that if a young man does not accept his parents' values, he must be actively rejecting them. And when the generations face similar life situations, emulation and rejection are indeed the two main possibilities. But rebellion, after all, presup-

171

poses that the target of one's hostility is an active threat: in classical stories of filial rebellion, the son is in real danger of being forced to become like his father, and he rebels rather than accept this definition of himself. But when a young man simply sees no possibility of becoming like his parents, then their world is so remote that it neither tempts nor threatens him. Indeed, many a youth is so distant from his parents, in generational terms if not in affection, that he can afford to "understand" them, and often to show a touching sympathy for their hesitant efforts to guide and advise him. Parents, too, often sense that they appear dated or "square" to their children; and this knowledge makes them the more unwilling to try to impose their own values or preferences. The result is frequently an unstated "gentleman's agreement" between the generations that neither will interfere with the other. This understanding acknowledges a real fact of existence today; but just as often, it creates new problems.

One of these problems appears very vividly in the *absence of paternal exemplars* in many contemporary plays, novels, and films. One of the characteristic facts about most of our modern heroes is that they have no fathers—or, when they do have fathers, these are portrayed as inadequate or in some other way as psychologically absent. Take Augie March or Holden Caulfield, take the heroes of Arthur Miller's and Tennessee Williams' plays, or consider the leading character in a film like *Rebel Without A Cause*. None of them has a father who can act as a model or for that matter as a target of overt rebellion. The same is true, though less dramatically, for a great many young people today. One sometimes even hears students in private conversations deplore the tolerance and permissiveness of their exemplary parents: "If only, just once, they would tell me what *they* think I should do." Young people want and need models and guardians of their development; and they usually feel cheated if they are not available. The gentleman's agreement seldom works.

It would be wrong, however, to infer that parents have suddenly become incompetent. On the contrary, most American parents are genuinely interested in their children, they try hard to understand and sympathize with them, they continually think and worry about how to guide their development. In other, more stable times, these same parents would have been excellent models for their children, nourishing their growth while recognizing their individuality. But today they often leave their children with a feeling of never really

172

having had parents, of being somehow cheated of their birthright. The explanation is not hard to find; even the most well-intentioned parent cannot now hope to be a complete exemplar for his children's future. A man born in the 1910's or 1920's and formed during the Depression already finds himself in a world that was inconceivable then; his children will live in a world still more inconceivable. It would be unrealistic to hope that they would model their lives on his.

Another aspect of the psychology of rapid change is the *widespread feeling of powerlessness*—social, political, and personal—of many young people today. In the 1930's, there was a vocal minority which believed that society should and, most important, *could* be radically transformed; and there were more who were at least convinced that their efforts mattered and might make a difference in politics and the organization of society. Today the feeling of powerlessness extends even beyond matters of political and social interest; many young people see themselves as unable to influence any but the most personal spheres of their lives. The world is seen as fluid and chaotic, individuals as victims of impersonal forces which they can seldom understand and never control. Students, for example, tend not only to have a highly negative view of the work of the average American adult, seeing it as sterile, empty, and unrewarding, but to feel themselves caught up in a system which they can neither change nor escape. They are pessimistic about their own chances of affecting or altering the great corporations, bureaucracies, and academies for which most of them will work, and equally pessimistic about the possibility of finding work outside the system that might be more meaningful.

Such feelings of powerlessness of course tend to be self-fulfilling. The young man who believes himself incapable of finding a job outside the bureaucratic system and, once in a job, unable to shape it so that it becomes more meaningful will usually end up exactly where he fears to be—in a meaningless job. Or, a generation which believes that it cannot influence social development will, by its consequent lack of involvement with social issues, in fact end up powerless before other forces, personal or impersonal, which *can* affect social change. In a generation as in individuals, the conviction of powerlessness begets the fact of powerlessness.[11] But, however incorrect, this conviction is easy to comprehend. The world has always been amazingly complex, and with our widening understanding comes a sometimes paralyzing awareness of its complexity. Furthermore, when one's vantage point is continually shifting, when

173

the future is in fact more changeable than ever before, when the past can provide all too few hints as to how to lead a meaningful life in a shifting society—then it is very difficult to sustain a conviction that one can master the environment.

The most common response to this feeling of helplessness is what David Riesman has called *privatism*. Younger people increasingly emphasize and value precisely those areas of their lives which are least involved in the wider society, and which therefore seem most manageable and controllable. Young men and women today want large families, they are prepared to work hard to make them good families, they often value family closeness above meaningful work, many expect that family life will be the most important aspect of their lives. Within one's own family one seems able to control the present and, within limits, to shape the future. Leisure, too, is far more under the individual's personal control than his public life is; a man may feel obliged to do empty work to earn a living, but he can spend his leisure as he likes. Many young people expect to find in leisure a measure of stability, enjoyment, and control which they would otherwise lack. Hence their emphasis on assuring leisure time, on spending their leisure to good advantage, on getting jobs with long vacations, and on living in areas where leisure can be well enjoyed. Indeed, some anticipate working at their leisure with a dedication that will be totally lacking in their work itself. In leisure, as in the family, young people hope to find some of the predictability and control that seem to them so absent in the wider society.

Closely related to the emphasis on the private spheres of life is the *foreshortening of time span*. Long-range endeavors and commitments seem increasingly problematical, for even if one could be sure there will be no world holocaust, the future direction of society seems almost equally uncertain. Similarly, as the past becomes more remote, in psychological terms if not in actual chronology, there is a greater tendency to disregard it altogether. The extreme form of this trend is found in the "beat" emphasis on present satisfactions, with an almost total refusal to consider future consequences or past commitments. Here the future and the past disappear completely, and the greatest possible intensification of the present is sought. In less psychopathic form, the same emphasis on pursuits which can be realized in the present for their own sake and not for some future reward is found in many young people. The promise of continuing inflation makes the concept of a nest egg obsolete, the guarantee of changing job markets makes commitment to a specialized skill problematical, the possibility of a war, if seriously entertained, makes all future planning ridiculous. The consequence is that only the rare young man has life goals that extend more than five or ten years ahead; most can see only as far as graduate school, and many simply drift into, rather than choose, their future careers. The long-range goals, postponed satisfactions, and indefinitely deferred rewards of the Protestant Ethic are being replaced by an often reluctant hedonism of the moment.

A corollary of the emphasis on the private and the present is the *decline in political involvement* among college youth. To be sure, American students have never evinced the intense political concerns of their Continental contemporaries, and admittedly, there are exceptions, especially in the "direct-action" movements centered around desegregation. But the general pattern of political disengagement remains relatively unchanged, or if anything has become more marked. Those familiar with elite college students in the 1930's and in the late 1950's contrast the political activity of a noisy minority then with the general apathy now before world problems of greater magnitude. Instead of political action, we have a burgeoning of the arts on many campuses, with hundreds of plays, operas, poems, and short stories produced annually by college students. Underlying this preference of aesthetic to political commitment are many of the outlooks I have mentioned: the feeling of public powerlessness, the emphasis on the private and immediate aspects of life, the feeling of disengagement from the values of the parental generation. But most important is the real anxiety that overtakes many thoughtful young people when they contemplate their own helplessness in the face of social and historical forces which may be taking the world to destruction. It is perhaps significant that Harvard students began rioting about Latin diplomas the evening of a relatively underattended rally to protest American intervention in Cuba, a protest to which most students would have subscribed. So high a level of anxiety is generated by any discussion of complex international relations, the possibilities of nuclear war, or even the complicated issues of American domestic policies, that all but the extraordinarily honest or the extraordinarily masochistic prefer to release their tensions in other ways than in political activity. And in this disinvolvement they are of course supported by the traditional American myth of youth, which makes it a time for panty raids but not for politics.

In general, then, many college students have a kind of *cult of*

175

experience, which stresses, in the words of one student, "the maximum possible number of sense experiences." Part of the fascination which the beat generation holds for college students lies in its quest for "kicks," for an intensification of present, private experiences without reference to other people, to social norms, to the past or the future. Few college students go this far, even in the small group that dresses "beat," rides motorcycles, and supports the espresso bars; for most, experience is sought in ways less asocial than sex, speed, and stimulants. But travel, artistic and expressive experience, the enjoyment of nature, the privacy of erotic love, or the company of friends occupy a similar place in the hierarchy of values. Parallel with this goes the search for self within the self rather than in society, activity or commitment, and a belief that truth can be uncovered by burrowing within the psyche. The experience sought is private, even solipsistic; it involves an indifference to the beckonings of the wider society. To be sure, Teddy Roosevelt, too, was in his way a seeker after experience; but unlike most contemporary American youths, he sought it in frantic extroversion, in bravado and heroic action; and its rewards were eventual public acclaim. But for most college students today, T.R. and the values of his era have become merely comic.

Youth Culture and Identity

Many of these outlooks of youth can be summed up as a sophisticated version of the almost unique American phenomenon of the "youth culture," [12] that is, the special culture of those who are between childhood and adulthood, a culture which differs from both that of the child and that of the adult. To understand the youth culture, we must consider not only the increasing gap between the generations but the discontinuity between childhood and adulthood. [13] Generational discontinuities are gaps in time, between one *mature* generation and the next; but age group discontinuities are gaps between different age groups at the *same* time. The transition from childhood to adulthood is never, in any society, completely continuous; but in some societies like our own there are radical discontinuities between the culturally given definitions of the child and of the adult. The child is seen as irresponsible, the adult responsible; the child is dependent, the adult is independent; the child is supposedly unsexual, the adult is interested in sex; the child plays, the adult works, etc. In societies where these age-group discontinui-

176

ties are sharpest, there is usually some form of initiation rite to guarantee that everyone grows up, that the transition be clearly marked, and that there be no backsliding to childish ways.

But in our society we lack formalized rites of initiation into adulthood; the wan vestiges of such rites, like bar mitzvah, confirmation, or graduation-day exercises, have lost most of their former significance. Instead, we have a youth culture, not so obviously transitional, but more like a waiting period, in which the youth is ostensibly preparing himself for adult responsibilities, but in which to adults he often seems to be armoring himself against them. Of course, the years of the youth culture are usually spent in acquiring an education, in high school, college, vocational or professional training. But it would be wrong to think of the youth culture as merely an apprenticeship, a way of teaching the young the technical skills of adulthood. For the essence of the youth culture is that it is not a rational transitional period—were it one, it would simply combine the values of both childhood and adulthood. Instead, it has roles, values, and ways of behaving all its own; it emphasizes disengagement from adult values, sexual attractiveness, daring, immediate pleasure, and comradeship in a way that is true neither of childhood nor of adulthood. The youth culture is not always or explicitly anti-adult, but it is belligerently *non*-adult. The rock'n' roller, the Joe College student, the juvenile delinquent, and the beatnik, whatever their important differences, all form part of this general youth culture.

To understand this subculture we must consider its relation to both the discontinuities between age groups and the discontinuities between generations. I have noted that young people frequently view the more public aspects of adult life as empty, meaningless, a rat race, a futile treadmill; only in private areas can meaning and warmth be found. Childhood contrasts sharply with this image: childhood is seen as (and often really is) a time for the full employment of one's talents and interest, a time when work, love, and play are integrally related, when imagination is given free play, and life has spontaneity, freedom, and warmth. Adulthood obviously suffers by comparison, and it is understandable that those who are being rushed to maturity should drag their feet if this is what they foresee. The youth culture provides a kind of way-station, a temporary stopover in which one can muster strength for the next harrowing stage of the trip. And for many, the youth culture is not merely one of the stops, but the last stop they will really enjoy or feel commitment to.

177

Thus, the youth culture is partially a consequence of the discontinuity of age groups, an expression of the reluctance of many young men and women to face the unknown perils of adulthood.

But the gap between childhood and adulthood will not explain why in our society at present the youth culture is becoming more and more important, why it involves a greater and greater part of young men and women's lives, or why it seems so tempting, compared with adulthood, that some young people increasingly refuse to make the transition at all. Rock'n'roll, for example, is probably the first music that has appealed almost exclusively to the youth culture; catering to the teenage market has become one of the nation's major industries. And, as Riesman has noted, the very word "teenager" has few of the connotations of transition and growing up of words like "youth" and "adolescent," which "teenager" is gradually replacing.[14]

The youth culture not only expresses youth's unwillingness to grow up, but serves a more positive function in resolving generational discontinuities. Erik H. Erikson would characterize our youth culture as a psychosocial moratorium on adulthood, which provides young people with an opportunity to develop their identity as adults.[15] One of the main psychological functions of a sense of identity is to provide a sense of inner self-sameness and continuity, to bind together the past, the present, and the future into a coherent whole; and the first task of adolescence and early adulthood is the achievement of identity. The word "achieve" is crucial here, for identity is not simply given by the society in which the adolescent lives; in many cases and in varying degrees, he must make his own unique synthesis of the often incompatible models, identifications, and ideals offered by society. The more incompatible the components from which the sense of identity must be built and the more uncertain the future for which one attempts to achieve identity, the more difficult the task becomes. If growing up were merely a matter of becoming "socialized," that is, of learning how to "fit into" society, it is hard to see how anyone could grow up at all in modern America, for the society into which young people will some day "fit" remains to be developed or even imagined. Oversimplifying, we might say that socialization is the main problem in a society where there are known and stable roles for children to fit into; but in a rapidly changing society like ours, identity formation increasingly replaces socialization in importance.

Even the achievement of identity, however, becomes more difficult in a time of rapid change. For, recall that one of the chief tasks of identity formation is the creation of a sense of self that will link the past, the present, and the future. When the generational past becomes ever more distant, and when the future is more and more unpredictable, such continuity requires more work, more creative effort. Furthermore, as Erikson emphasizes, another of the chief tasks of identity formation is the development of an "ideology," that is, of a philosophy of life, a basic outlook on the world which can orient one's actions in adult life. In a time of rapid ideological change, it seldom suffices for a young man or woman simply to accept some ideology from the past. The task is more difficult; it involves selecting from many ideologies those essential elements which are most relevant and most enduring. Such an achievement takes time, and sometimes the longest time for the most talented, who usually take the job most seriously.

The youth culture, then, provides not only an opportunity to postpone adulthood, but also a more positive chance to develop a sense of identity which will resolve the discontinuity between childhood and adulthood on the one hand, and bridge the gap between the generations on the other. Of course, a few young men and women attempt to find an alternative to identity in other-direction. Unable to discover or create any solid internal basis for their lives, they become hyperadaptable; they develop extraordinary sensitivity to the wishes and expectations of others; in a real sense, they let themselves be defined by the demands of their environment. Thus, they are safe from disappointment, for having made no bets on the future at all, they never have put their money on the wrong horse. But this alternative is an evasion, not a solution, of the problem of identity. The other-directed man is left internally empty; he has settled for playing the roles that others demand of him. And role-playing does not satisfy or fulfill; letting the environment call the shots means having nothing of one's own. Most young people see this very clearly, and only a few are tempted to give up the struggle.

There is another small group, the so-called beats and their close fellow-travelers, who choose the other alternative, to opt out of the System altogether and to try to remain permanently within the youth culture. In so doing, some young people are able to create for themselves a world of immediate, private and simple enjoyment. But leaving the System also has its problems. The search for self which runs through the youth culture and the beat world is not the whole of life, and to continue it indefinitely means usually renouncing attainments which have been traditionally part of the definition of a man or a woman: intimacy and love for others; personal creativity in work, ideas, and children; and that fullness and roundedness of life which is ideally the reward of old age. So, though many young people are tempted and fascinated by the beat alternative, few actually choose it.

The vast majority of young people today accept neither the other-directed nor the beat evasion of the problem of identity. In many ways uncommitted to the public aspects of adult life, they are willing nonetheless to go through the motions without complete commitment. They have a kind of "double consciousness," one part oriented to the adult world which they will soon enter, the other part geared to their version of the youth culture. They are not rebellious (in fact they like their parents), but they feel estranged and distant from what their elders represent. They often wish they could model themselves after (or against) what their parents stand for, but they are sensible enough to see that older people are often genuinely confused themselves. They feel relatively powerless to control or to influence the personal world around them, but they try to make up for this feeling by emphasizing those private aspects of life in which some measure of predictability and warmth can be made to obtain. They often take enthusiastic part in the youth culture, but most of them are nonetheless attempting to "graduate" into adulthood. And though many hesitate on the threshold of adulthood, they do so not simply from antagonism or fear, but often from awareness that they have yet to develop a viable identity which will provide continuity both within their lives and between their own, their parents', and their future children's generations. And in each of these complex and ambivalent reactions young people are in part responding to the very process of unrestrained change in which they, like all of us, are involved.

Evaluations and Prospects

In these comments so far I have emphasized those attitudes which seem most directly related to the stresses of unrestrained change, neglecting other causal factors and painting a somewhat dark picture. I have done this partly because the more sanguine view of youth—which stresses the emancipations, the sociological understandability of youth's behavior, the stability of our society despite unprecedented changes, and the "adaptive" nature of youth's behavior—this more encouraging view has already been well presented.[16] But furthermore, if we shift from a sociological to a psychological perspective and ask how young people themselves experience growing up in this changing society, a less hopeful picture emerges. Rightly or wrongly, many young people experience emancipations as alienations; they find their many freedoms burdensome without criteria by which to choose among equally attractive alternatives; they resent being "understood" either sociologically or psychologically; and they often find the impressive stability of our society either oppressive or uninteresting. Furthermore, what may constitute an "adaptation" from one sociological point of view (e.g., the American Indian's regression in the face of American core culture) may be not only painful to the individual but disastrous to the society in the long run. A sociological and a psychological account of youth thus give different though perhaps complementary pictures, and lead to different evaluations of the outlook of American youth. Despite the stability of American society and the undeniable surfeit of opportunities and freedoms available to young people today, many of youth's attitudes seem to me to offer little ground for optimism.

The drift of American youth, I have argued, is away from public involvements and social responsibilities and toward a world of private and personal satisfactions. Almost all young people will eventually be *in* the system—that is, they will occupy occupational and other roles within the social structure—but a relatively large number of them will never be *for* the system. Like the stereotypical Madison Avenue ad-man who works to make money so that he can nourish his private (and forever unrealized) dream of writing a novel, their work and their participation in public life will always have a somewhat half-hearted quality, for their enthusiasms will be elsewhere—with the family, the home workshop, the forthcoming vacation, or the unpainted paintings. Their vision and their consciousness will be split, with one eye on the main chance and the other eye (the better one) on some private utopia. This will make them good organizational workers, who labor with detachment and correctness but without the intensity or involvement which might upset bureaucratic applecarts. And they will assure a highly stable political and social order, for few of them will be enough committed to politics to consider revolution, subversion, or even radical change. This orientation also has much to commend it to the individual: the private and immediate is indeed that sphere subject to the greatest

181

personal control, and great satisfaction can be found in it. The "rich full life" has many virtues, especially when contrasted with the puritanical and future-oriented acquisitiveness of earlier American generations. And I doubt if commitment and "fidelity" will disappear; rather, they will simply be transferred to the aesthetic, the sensual, and the experiential, a transfer which would bode well for the future of the arts.

Yet the difficulties in this split consciousness seem to me overwhelming, both for the individual and for the society. For one, few individuals can successfully maintain such an outlook. The man who spends his working day at a job whose primary meaning is merely to earn enough money to enable him to enjoy the rest of his time can seldom really enjoy his leisure, his family, or his avocations. Life is of a piece, and if work is empty or routine, the rest will inevitably become contaminated as well, becoming a compulsive escape or a driven effort to compensate for the absent satisfactions that should inhere in work. Similarly, to try to avoid social and political problems by cultivating one's garden can at best be only partly successful. When the effects of government and society are so ubiquitous, one can escape them only in the backwaters, and then only for a short while. Putting work, society, and politics into one pigeonhole, and family, leisure and enjoyment into another creates a compartmentalization which is in continual danger of collapsing. Or, put more precisely, such a division of life into nonoverlapping spheres merely creates a new psychological strain, the almost impossible strain of artificially maintaining a continually split outlook.

Also on the demerit side, psychologically, is the willful limitation of vision which privatism involves, the motivated denial of the reality or importance of the nonprivate world. Given the unabating impact of social forces on every individual, to pretend that these do not exist (or that, if they do exist, have no effect on one) qualifies as a gross distortion of reality. Such blindness is of course understandable: given the anxiety one must inevitably feel before a volatile world situation, coupled with the felt inability to affect world events, blinders seem in the short run the best way to avoid constant uneasiness. Or similarly, given the widespread belief that work is simply a way of earning a living, refusal to admit the real importance to one's psychic life of the way one spends one's working days may be a kind of pseudo-solution. But a pseudo-solution it is, for the ability to acknowledge unpleasant reality and live with the attendant anxiety is one of the criteria of psychological health. From a psychological point of view, alienation and privatism can hardly be considered ideal responses to social change.

From a social point of view, the long-range limitations of these "adaptations" seem equally great. Indeed, it may be that, through withdrawal from concern with the general shape of society, we obtain short-run social stability at the price of long-run stagnation and inability to adapt. Young people, by exaggerating their own powerlessness, see the "system," whether at work, in politics, or in international affairs, as far more inexorable and unmalleable than it really is. Consider, for example, the attitude of most American youth (and most older people as well) toward efforts to direct or restrain the effects of social change. Partly by a false equation of Stalinism with social planning, partly on the assumption that unrestrained social change is "natural," and partly from a conviction that social planning is in any case impossible, young people usually declare their lack of interest. Apart from the incorrectness of such beliefs, their difficulty is that they tend to be self-confirming in practice. Given a generation with such assumptions, social changes will inevitably continue to occur in their present haphazard and unguided way, often regardless of the needs of the public. Or again, it seems likely that if any considerable proportion of American students were to demand that their future work be personally challenging and socially useful, they would be able to create or find such work and would revolutionize the quality of work for their fellows in the process. But few make such demands. Or, most ominous of all, if the future leaders of public opinion decide that they can leave the planning of foreign policy to weapons experts and military specialists, there is an all too great chance that the tough-minded "realism" of the experts will remain unmitigated by the public's wish to survive.

In short, an alienated generation seems too great a luxury in the 1960's. To cultivate one's garden is a stance most appropriate to times of peace and calm, and least apposite to an era of desperate international crisis. It would be a happier world than this in which men could devote themselves to personal allegiances and private utopias. But it is not this world. International problems alone are so pressing that for any proportion of the ablest college students to take an apolitical stance seems almost suicidal. And even if world problems were less horrendous, there is a great deal to be done in our own society, which to many, young and old, still seems corrupt, unjust, ugly, and inhuman. But to the extent that the younger generation loses interest in these public tasks, remaining content with private virtue, the public tasks will remain undone. Only a utopia can afford alienation.

In so far as alienation and privatism are dominant responses of the current college generation to the stresses of unrestrained change, the prospects are not bright. But for several reasons, I think this prognosis needs qualification. For one, I have obviously omitted the many exceptions to the picture I have sketched—the young men and women who have the courage to confront the problems of their society and the world, who have achieved a sense of identity which enables them to remain involved in and committed to the solution of these problems. Furthermore, for most students alienation is a kind of *faute de mieux* response, which they would readily abandon, could they find styles of life more deserving of allegiance. Indeed, I think most thoughtful students agree with my strictures against privatism, and accept withdrawal only as a last resort when other options have failed. But, most important, I have omitted from my account so far any discussion of those forces which do or might provide a greater sense of continuity, despite rapid change. Discussion of these forces may correct this perhaps unnecessarily discouraged picture.

Throughout this account, I have suggested that Americans are unwilling to plan, guide, restrain, or coordinate social change for the public good. While this is true when America is compared with other industrialized nations, it is less true than in the past, and there are signs that many Americans are increasingly skeptical of the notion that unrestrained change is somehow more "free" or more "natural" than social planning. We may be beginning to realize that the decision not to plan social changes is really a decision to allow forces and pressures other than the public interest to plot the course of change. For example, it is surely not more natural to allow our cities to be overrun and destroyed by the technological requirements of automobiles than to ask whether humane and social considerations might not require the banning or limiting of cars in major cities. Or to allow television and radio programming to be controlled by the decisions of sponsors and networks seems to many less "free" than to control them by public agencies. If we are prepared to guide and limit the course of social change, giving a push here and a pull there when the "natural" changes in our society conflict with the needs of the public, then the future may be a less uncertain prospect for our children. Indeed, if but a small proportion of the energy we now spend in trying to second-guess the future were channelled into efforts to shape it, we and our children might have an easier task in discovering how to make sense in, and of, our changing society.

I have also neglected the role that an understanding of their situation might play for the younger generation. Here I obviously do not mean that students should be moralistically lectured about the need for social responsibility and the perversity of withdrawal into private life. Such sermonizing would clearly have the opposite effect, if only because most young people are already perfectly willing to abandon privatism if they can find something better. But I do mean that thoughtful students should be encouraged to understand the meaning and importance of their own stage in life and of the problems which affect them as a generation. The emphasis on individual psychological understanding which characterizes many "progressive" colleges can provide only a part of the needed insight. The rest must come from an effort to end the pluralistic ignorance of the stresses confronting all members of the current younger generation. Here colleges do far too little, for courses dealing with the broad social pressures that impinge on the individual often deliberately attempt to prevent that personal involvement which alone gives insight. But one can imagine that a concrete understanding of the psychosocial forces that affect a generation might have some of the same therapeutic effects on the more reflective members of the generation that insight into psychodynamic forces can give the thoughtful individual.

And finally, I have underplayed the importance that values and principles can and do play in providing continuity amid rapid change. If one is convinced that there are guiding principles which will remain constant—and if one can find these enduring values—life can be meaningful and livable despite rapid change. But here we need to proceed cautiously. Technologies, institutions, ideologies, and people—all react by extremes when faced with the fear of obsolescence. Either they firmly insist that *nothing* has changed and that they are as integrally valid as ever before or—and this is equally disastrous—they become so eager to abandon the outmoded that they abandon essential principles along with the irrelevant. Thus, parents who dimly fear that they may appear "square" to their children can react either by a complete refusal to admit that anything has changed since their early days or (more often) by suppressing any expression of moral concern. The second alternative seems to me the more prevalent and dangerous. An antiquated outlook is

185

usually simply ignored by the young. But person or institution that abandons its essential principles indirectly communicates that there are no principles which can withstand the test of time, and thus makes the task of the young more difficult.

Yet the bases for the continuity of the generations must of necessity shift. Parents can no longer hope to be literal models for their children; institutions cannot hope to persist without change in rite, practice, and custom. And, although many of the essential principles of parents, elders, and traditional institutions can persist, even those who seek to maintain the continuity of a tradition must, paradoxically, assume a creative and innovating role. We need not only a rediscovery of the vital ideals of the past, but a willingness to create new ideals—new values, new myths, and new utopias—which will help us to adapt creatively to a world undergoing continual and sweeping transformations. It is for such ideals that young people are searching: they need foundations for their lives which will link them to their personal and communal pasts and to their present society but which at the same time will provide a trustworthy basis for their futures. The total emulation or total rejection of the older generation by the young must be replaced by a recreation in each generation of the living and relevant aspects of the past, and by the creation of new images of life which will provide points of constancy in a time of rapid change.

REFERENCES

1. An earlier version of parts of this paper was presented at the Annual Conference of Jewish Communal Services, May 1961, and was published in *The Journal of Jewish Communal Services* (Fall 1961).

2. It need hardly be added that our society's capacity for innovation and change is also one of its greatest strengths.

3. Among the other major factors creating stresses for American youth are (1) the discontinuities between childhood and adulthood, especially in the areas of sex, work, and dependency; (2) the great rise in the aspirations and standards of youth, which create new dissatisfactions; and (3) the general intellectual climate of skepticism and debunking, which makes "ideological" commitment difficult. In this essay, however, I will concentrate on the stresses created by social change.

4. One should not confuse static with stable societies. American society is extremely stable internally despite rapid rates of change. Similarly, other societies, though relatively static, are unstable internally.

5. Unconsciously, however, most Americans have highly ambivalent feelings about science and technology, usually expressed in the myth of the (mad) scientist whose creation eventually destroys him.

6. See Walter Rosenblith, "On Some Social Consequences of Scientific and Technological Change," *Dædalus* (Summer 1961), pp. 498-513.

7. Obviously, the existence of institutions and values opposed to technological change in a technological society is itself a major source of social and individual tension.

8. Other types of social change also have their own characteristic world views. In particular, the mentality of elite youth in underdeveloped countries now beginning industrialization differs from that in transitional countries like Japan, where technological and pretechnological elements coexist. American society probably comes closest to a "pure" type of exclusively technological change.

9. Once again I omit any discussion of other sources of strain on youth (see reference 3). Furthermore, I do not mean to suggest that these outlooks are the only possible responses to unrestrained change, or that they are unaffected by other historical and social forces in American life.

10. Paul Goodman, *Growing Up Absurd*. New York: Random House, 1960.

11. It is ironic that this generation, which is better prepared than any before it, which knows more about itself and the world and is thus in a better position to find those points of leverage from which things can be changed, should feel unable to shape its own destiny in any public respect.

12. Talcott Parsons, "Age and Sex Grading in the United States," reprinted in Parsons, *Essays in Sociological Theory, Pure and Applied* (Glencoe, Illinois: The Free Press, 1949). The beginnings of a youth culture are appearing in other highly industrialized countries, which suggests that this institution is characteristic of a high degree of industrialization.

13. Ruth Benedict, "Continuities and Discontinuities in Cultural Conditioning," in Clyde Kluckhohn and Henry A. Murray (eds.), *Personality in Nature, Society, and Culture*. New York: Norton, 1948.

14. David Riesman, "Where is the College Generation Headed?" *Harper's Magazine*, April 1961.

15. Erik H. Erikson, "The Problem of Ego Identity," in *Identity and the Life Cycle*, published as vol. I, no. 1 of *Psychological Issues* (1959). See also his "Youth: Fidelity and Diversity," in this issue.

16. Talcott Parsons, "Youth in the Context of American Society," in this issue.

ROBERT COLES

Serpents and Doves: Non-Violent Youth in the South

THERE ARE IN AMERICA today young men and women, black and white, who are going to jail for the freedom of their fellow men. They are doing radical things in novel and challenging ways; and they are doing them in every man's sight, in restaurants, in stores, in movie houses, in bus terminals, and in the obscure rural offices of voting registrars. In fact, wherever they are and wherever they go, they test and defy. Here, certainly for the first time in my life, I have seen American students behave toward their society with the "ideological" concern said to possess young people in Europe or Asia. But, they do not march with banners nor man barricades. They do not scream at other countries and their visiting leaders. They throw no bombs. They are concerned with their own country, and they want to become her true citizens. They very definitely want to change social and political customs, but they want to change them peacefully and in their time. They are not lost or confused; they know exactly what they want, and they are ready to give their lives for their goals. I went to study them and I came to respect them; and so I will tell their story so as to let them come to word, and through them, their tasks and their fate.

But how about the others, the men of business and the ordinary

The work upon which this paper rests is part of a study, now in its second year, concerned with the psychological problems among Negroes and whites which occur in desegregation, particularly in the schools. This effort is sponsored by the Southern Regional Council, a distinguished group of white and Negro Southerners who wish "to attain the ideals and practices of equal opportunity for all peoples in the South." I am deeply grateful to them, and wish to thank them and their fine staff. I especially wish to express my gratitude to the Executive Director, Mr. Leslie W. Dunbar, for his kindness and extraordinary intelligence.

188

citizens of all ages and classes who come up against these students? These people have grown up with certain deeply felt assumptions about ways of getting along with one another. This happens when you grow up anywhere—you learn who you are, and who you are not. You learn who is nice and who is bad. You learn who has rights and who doesn't seem to have any. Until recently, if you were born in the South and you were white, you learned that the colored man could not go certain places, could not have certain jobs, could not expect certain rights. You learned that he took care of you, did a lot of hard and difficult things for you. You were told about his bondage, and soon learned about the Civil War. Associated in your mind with the Negro were the hard times of the South after that war, the pride and power of the North, and your unending conviction, based on the laws of human nature, that there was hypocrisy under all those words from up there.

These traditions and sentiments are not tidily bound in a cellophane bag. They are carried around by all those human beings in Greensboro, North Carolina, Atlanta, Georgia, and every other town in the South. If you look at those towns in 1962, you will see television sets and air fields. Living in these towns are young people who expect someone of their generation to land on the moon. Perhaps they think of themselves as Southerners; but they also see themselves as living on a patch of a planet which is orbited, radioactive, and supersonically traveled. We all know how their parents struggled over the question of how or whether America should get involved in the problems of distant continents. Somehow we did become involved; somehow the distance between the continents has been abolished. Now we worry about countries who want to bury us and continents where new countries are rising from the ashes of old empires. We worry, and we gird ourselves for a long haul of trouble. We pull ourselves together, and parts of us get closer together. Factories move from one region to another, taking people with them, drawing people to them. Farms are left. Cities grow. People leave places where they can't work or feel hunted and go to other places where they still have their troubles but where these seem less overwhelming. Feelings about dark people persist, but must now be put on complicated scales and weighed with other feelings about work and country. Feelings about white people persist, but the white world is changing and so is the dark, and new feelings arise and demand expression. The towns and cities hold all that, grapple with paradox and contradiction and new ideas and new events; they will live or

189

die on whether they settle in themselves how they get along with themselves.

A Southern journalist emphasizes, with a newsman's special pleasure in irony, that the four young Negroes who first dared, in 1960, to ask for dimestore coffee in Greensboro, North Carolina, were prompted, even pushed, by whites impatient with the pace of the Negro's walk to full citizenship and pressing for changes. Apocryphal or not, there is truth in the story. The truth is that students are attacking conventions by picketing and by boycotts; they are going on freedom rides and risking arrest for their sit-ins. The additional truth is that when they lean hard, the walls, more often than not, crumble. It takes a certain kind of society for this to happen. In some countries the walls would not crumble. The protests would be crushed, turned into revolutionary plots. In our own country such protests have been made many times in the past with little success. There were sit-ins decades ago, which are remembered now because the present ones work. Because they work, they say something about the country and about the South. Social changes occur when a ripe moment is plucked by shrewd and hungry men. This ripe moment in the South today captures the world's interest. One of the great issues in this century is whether the different races of the world will be able to get along in some new mutual respect. The alternative is grim, but may help make it possible. The world watches a great protagonist try to make its ideals real. The South is the troubled stage, and much depends upon the resolution of the drama.

There are many actors. Some of them are storekeepers, some are governors, some are policemen. But is is students who seem to initiate much of the action and cause many of the changes; and we are here to try to look at these lean and crafty youths. We might ask ourselves about their aims and their tasks, their ages and origins, their manner of living and the ways they respond to some of the troubles thrown their way.

What they want is called "desegregation," or "integration." Some call it "mixing." Under the names are wishes and fears. The students wish that the Negro people in the South could have an easier, kinder life. They would like a Negro to feel free to vote. They would like him to a hold a job which bears some connection with his interests or abilities. The time has come, they assert, when a Negro can eat in those restaurants and sit in those movies. Parks are for people, and they want to feel like people. No other group of Americans has been shunted aside so long, kept in such degradation, exploited so

190

unremittingly. The students say all these words, and that all bad things must come to an end, and that the glorious moment has come. They seem to have the idea that a secret voice has been heard and read by them. This secret voice tells them that their country can no longer endure with the old segregated ways. It goes on to suggest that the past years have shown a slow yielding of the law and a gradual improvement in the separate but equal position of the Negro. Then it bellows into their ears and scrawls on their student notebooks the advice that to freedom there are many roads. Their road is that of the direct and non-violent protest. They are young, and they feel directly as the young do; they direct their attention to some very specific customs, and they say to one another that these exist in the middle of the twentieth century only by their own sufferance. They sense that the South as well as their country is in a bind. Both would like to be rid of this shameful relic from other years. Both feel guilty, but both are afraid, and don't quite know how. They also sense that they have not been terrorized alone; that with them have been white people in every city and state of the South, people afraid to say what they feel and for this reason not free.

Their aim is not to conspire or to destroy their country. They want what others in the country have, what others have often taken for granted. They want what their country says it is all about. They want to feel they are more a part of their country. Then, as they let their minds wander into the future, they hope for more trust and friendliness between the white and dark people of this country. They carefully distinguish between desegregation and integration. They know that you cannot legislate affection. But they suspect that if many of their exiles end, hands of acceptance may begin to reach. One of their buttons shows a white hand and a black hand clasped. I suppose that this is their wish.

What they do to get what they want is to protest. They do not carry weapons. They do not hit or hurt people, at least physically. They will not fight with sticks or stones or hands or feet. They wear suits, and they shave. They are determined to appear as presentable young Americans. Before they protest they study and think about the way in which they will protest, and they practice this too. This is the way of non-violence. The way of non-violence is self-explanatory. But like many simple matters, it has complicated results, for the students and for the world they protest against.

It is complicated for them because they must control carefully some very strong feelings. It is in the nature of protest to be unhappy

191

over something and to be willing to assert that unhappiness. Mild unhappiness may be endured and not protested. Overwhelming unhappiness and indignity may also be endured and not protested, when protest is impossible because it will be crushed forthrightly. Such was the case in past years, though protests there were, and crushed they were. Now it is different. It is fairly clear that protest is possible, and changes inevitable. But the emotions which have been felt in the past, risen out of humiliation and injury, do not disappear. They urge protest, but they are not quiet emotions. They are emotions of anger and frustration and fear. Worse, they are those of shame and self-doubt. It is hard to deal with these emotions when people shout at you and attack you. When people remind you of all the lacerations and insults, it is hard to keep your cheek turned. You need the help of others, the help of rehearsals, and the help of some of the books which you left to make these protests. This idea has its roots in Christ's teachings. When they had very little else, Negroes were given the Christian religion. Insofar as the student has a tradition, insofar as his race has been allowed to develop one, it contains large amounts of this view. It is a tradition of bearing pain without protest. Then there is Gandhi, who studied Christ, and loved Tolstoy, and wanted freedom for his people, and didn't want to use violence to get it, and knew how to make it very embarrassing for those who wouldn't give it, and who got it.

Perhaps we should turn briefly to those protested against. Perhaps they will help us see how the students manage the balance between getting angry and controlling their anger. It is just as complicated for the white people. Most people are hurt when others are hurt. Most people are moved by what they see and hear. If the Negro is off yonder in some shabby slum, he can be forgotten, as he is very often in the North. If he is off yonder in some field by day and then in some shack by night, he can be forgotten, as he was in the South. Some very kindly and decent people in the South, for whom he has worked and labored, have never heard him complain, and have grown up with their special ways of talking with him and caring for him. It has not all been hate. But it is a kind of love, a kind of relationship, which is no longer wanted. Nor will the hate and insults be allowed anymore. This has to be told to many people. It is important to remember that many people simply do not know of this dissatisfaction. If they did, they would not repeat so often that the Negro would be happy were it not for sinister outside forces. These are the uninformed. There are also those who really are informed and have ached for years with the injustice which they see and feel. Such white Southerners, and they are legion, have known that when others suffer, they suffer, and that the lessons their children learn are the lessons of hate and confusion, the lesson that Christian messages do not apply to large groups of human beings, not really. All of these people will be forcefully confronted. So will many frightened Negro people, unable, unwilling, afraid to do as their brothers do.

Those who keep these protests happening are of many ages and cities and families. They are mostly black men, but there are many black women, and many white men and white women. Some are from poor families, some from those in better circumstances. The son of a distinguished Negro college president works beside a very poor and orphaned boy from a rural town. There are many kinds of demonstration, and they may draw upon different participants. Some are spontaneous and astonishingly successful. In Asheville, North Carolina, some boys and girls who were going to junior high school and high school decided one day to use the library, and to try to get sodas from several stores. That did it, and that surprised the Negro community perhaps more than the white. Some are spontaneous, but need much more planned and persistent action at later times for success. In Nashville many months were required. In Atlanta picketing of stores was followed by a boycott and long negotiation. More recently in Atlanta the movies houses were cautiously desegregated under an arrangement between the students and the owners, with the mayor acting as a friendly helper. But there are other demonstrations which are attempted in those desperate areas of last defiance, requiring careful action by experienced people. To dare to work in a small Mississippi town with Negro families, encouraging them to want to vote, tutoring them in their letters, helping them in the incredible intricacies of laws designed chiefly against them, requires more than a college weekend of time and experience. Some students stay in school and help during vacations or free time. Others leave school and devote months, years, to fulltime training and action. Most protests have their origin in the local town, the local Negro college, or one nearby. The freedom rides were an obvious exception. These rides were a thrust into the heart of segregation, the bus terminals in Alabama and Mississippi. They were followed by some removal of separate rest room signs. What they illustrated is the anxiety and concern which this country has for the welfare of these students. A policeman told one of them, "Even ten years ago you'd have been dead, all of you. Can't do that now." Though they are not sure, the students are sure enough to feel able to go ahead.

Those who go ahead seem to elude classification by class or geography or even age. College students predominate; but there have been younger boys and girls, and young married couples with children by their sides, and older people. In south Georgia or in Alabama or Mississippi or Louisiana any protest is made against heavy odds. In some towns in these states the mere act of protest is almost incredible. It is in these areas that violence is almost certain. It is here that terror stalks and danger is reliable. Police dogs are threatened in one state, and a man running for governor says that his body in front of the buildings will prevent execution of a federal law on school desegregation. He is elected. In another state students are arrested on charges of "criminal anarchy" for organizing sit-ins for coffee or encouraging campaigns to register voters. High bail is set, and some stay in prison and solitary confinement for months. In another state students escort some Negroes to register to vote. Crowds heckle and attack, arrests are made, students are kicked and hurt. But something prevents their murder, their lynching as in olden days. No one can say that such restraint in those towns is not some kind of communication from an outside world no longer so far away, so indifferent. The students leave jail, go back to their work.

Those students who leave school for some months and venture into these troublesome sections, live different lives once there. If you know the cramped, desolate, dreary quality to the Negro sections of Southern towns, if you've heard the spirituals and seen the funerals, diagnosed the tuberculosis and vitamin deficiencies, smelled the cheap, bad booze which blots out poverty, persecution, the whim of the alien white lords, then you will know how some of these students live. They will often come to one of these towns and live there for months, getting to know people and finding out the usual problems of these people and suggesting new and direct responses to these problems, like a boycott, or a school where people can be helped in reading and interpreting constitutional law so that they may try to register to vote. It is not simply the hardness of the whites, many of whom are very poor and very much struggling themselves, very soft and tender people, friendly and generous with the little they have, whose voices are honey and whose eyes are sly with earthly, human affairs and constricted with money troubles and job troubles and bitter old times which are getting better, and whose ancient consolation following ancient frustration was a lowly, yelping, always handy nigger. It is the apathy of the Negroes which is more difficult to look at and most painful of all. It is their inertia and resignation which are so formidable. Death of spirit or soul, by murder, by suicide, by illness and exploitation, is worse than live, howling, frightened enemies.

Some students stay in one town for months and help local leaders plan action. Others move about from town to town, state to state, talking with young Negro students or helping them form a plan. White students will go with them, or will go to white campuses, where they will contact those who are quietly, and in ways different from those in Negro colleges, working for changes. These students are often called "field secretaries." This means that they travel in the field of Southern cities and try to change that field by participating in actions which directly disregard its rules and boundaries. They receive enough money to travel and eat and sleep. When they come into a town they will look at the restaurants and the drugstore fountains. They will find out who votes, and where. They will watch parks and stand near libraries, and note who works at what job in those places where people work. Then they will get a room in some YMCA or bunk in some home or college dormitory, and start "making the rounds," hearing the many conflicting views of people, and recording and annotating them in their more than secretarial minds. After this, decisions must be made about what to do, in what order, and with whom.

When they decide, they decide their own fate, establish their own fears and hopes. Heavy odds they will be arrested and go to jail. Strong chance they will be kicked, punched, spat upon. Foul language is routine. Serious injury is possible. Most of these students have been in jail many times and in many states. Many are juggling several trial dates and bailed-free periods. They will go to court and be sentenced and appeal and, if there is money, go free and continue their activities, then interrupt them to go on trial in another state. Since most of what they do is illegal, and since they keep on in spite of this, they accumulate long prison records. Their fate is notorious. They are noticed by the press and television and radio. Their names may be given, their pictures taken. Leading newspapers and magazines will describe their behavior and, of course, its implications. This is important, because it is such news and widely spread consideration of this news which causes people to pay attention and become ashamed and indignant. All this is pressure on segregation, which, like many habits, works best unnoticed. When the students are noticed, they have scored a victory.

There are victories in the outside world and victories inside the mind and spirit of a person. Fears and hopes contend in us. Consider them both in some of these young people. They spend their time asking for things which they have been told they may not have; this is unpleasant enough. For asking, moreover, they are called criminals, and arrested, arraigned, imprisoned. This is unpleasant enough. As they prepare to ask, and while they ask, moreover, they risk injury and sustain insult. This is unpleasant enough. But when all is done, they must realize that what they have done is a very small beginning, and that can be very discouraging. And so there must be hope, and the mind must find ways to persist and continue when hope is sparse. It is in the nature of man for this to happen, and it does happen with these students. But not without some shifting and settling of emotions.

Abandoned are familiar routines and the comfort of daily rituals which are certain and predictable. Their clothes may be scattered over several states. There is no one bed and there may be many toothbrushes. These are students who have been brought up in homes where the discipline and routines of study are learned. In our civilization we cherish possessions and order. We learn in those early years which bed is our bed and which glass is our glass. We learn what time school opens and when it closes and how many hours we go. We learn numbers for our houses and numbers for our ages and numbers for our grades in school. We learn our private names and learn to put our writing on our books. We learn the order of classroom seating plans and graduation walking plans. We get scored in tests and ranked in classes, and we have certain pages to read for certain examinations. We all laugh when we go into the army and receive a long number beside our name; but the uneasy laugh is all too close to the long, familiar truth of our lives. When students leave college to risk punishment, leave a dormitory and classroom and library to face uncertain and changing quarters and work, they must make this possible for themselves. They are leaving what they have been brought up to value. They are leaving school, and the logical steps which they have been told to take as they become older; or they are staying in school but flirting with departure by their private, spare time actions. They are no longer listening to their parents, or, if their parents agree with what they do, they still have the memory of the older words of their parents, words suggesting study, success, accumulation of knowledge, books, ideas, degrees. Obviously, what each student does is balance in his own mind what he wants and what he wants to do. We may be interested in how these people think and what they feel. Each of them is unique, and personal reasons mesh with public activity with different results of happiness, nervousness, fearfulness. If we are going to know how the students get along with themselves or how they get along with others, we must talk with individual students over a long enough time to get some ideas and feelings about them. All the field secretaries live almost vagabond lives, but each field secretary comes to this life with his own resources and handicaps, and for his own reasons. We can watch people live and see them act, and record that. Then we can talk with them and listen to them and hear their words and emotions, and record that, too.

Here is some information about an American man of twenty who left college in his sophomore year to engage in some of these protesting actions. Let him be called John.

He was born in a small farming town of southern Virginia in the summer of 1942, shortly after we went to war. His father was away fighting in Europe during his first three years. This was the first time his father had ever left his home. His mother worked in the home of a white lawyer, taking care of three children. John was left to his maternal grandmother. He recalls her as his childhood mother, and talks about her even now with much feeling. He remembers his father's return, and the feeling of awkwardness and strangeness. A younger brother was soon to come. His father had trouble finding a job, and seemed irritable and punishing to the boy. He remembers being beaten with a strap, and running away. He remembers being punished when about five for trying to go to a circus, trying to escape his parents and enter and go on the merry-go-round. His grandmother must have told him many times about the difference between white and colored children, but he recalls that he "really learned it" when he tried to sit beside a white lady one day on a bus.

I used to straggle after my grandmother anyway, and I must have done it on the bus. She kept on pushing to the rear, and it didn't seem to have any seats. I just sat down in front. The woman pushed me and told me to get away and go to the back. I can remember how I felt then. I can still feel that feeling, like the world collapsing and not being liked. I felt I was bad and different, and I can still see that bus and that woman's look. Then, I was lectured by my grandmother, and her words drove everything I'd heard home for good. But my father was not happy with that, and he gave me the worst beating of my life when I got home. I can

197

recall that too; I'll never forget him . . . telling me to keep my place. . . . My grandmother cried and so did my mother.

He tells about his school work. He was a good student, and the teachers told him that he should stay in school. Most of his friends left school when about twelve or thirteen, going to the fields. His mother would tell him about big and mysterious law books in her home-by-day. She would come home in the evening and tell the family about all the events in "the boss house." He recalls his fantasies about this, and the pleasure which those stories brought to the poor home. His father had by now a job in a new factory which came south after the war. The boy went to high school, and this was quite a distinction; in the entire family, spread over the nearby towns, he was the first. At fourteen he fell in love with a girl a year older, who admired him "for my brains and promise." He wanted to leave school and marry her and get a job and she wanted him to stay at studying. They fought; he stayed; she left. He graduated and was awarded a scholarship to a Negro college far enough away to be his break with home. He left, a family hero and with great promise. "I remember," he tells in his deep and strong voice:

I can *still* remember reading about a sit-in which fellows from college were staging when I came there . . . I can *still* remember the Little Rock trouble. I used to have dreams about that. We'd sit and watch that on that television set, and we'd get angry, and my grandmother would say that those kids shouldn't try to go to that school, and my mother wouldn't say anything, and my father would curse those whites and say we should stay away from *all* of them . . . they can't be trusted, they're no good. And my mother would tell me, sometimes in a corridor away from dad, that that wasn't so, and then I'd hear about the lawyer and his family. . . . Mother told me one day that the lawyer's oldest son told her that he was convinced that the South was changing, and that he didn't mind going to school with Negroes . . . Mother told me that he used the word "Negro" . . . she can just tell who thinks what about us in that house . . . like when they stop talking about something when she comes into the room. You can tell . . . I'd go to bed and dream that I was one of those nine kids in Central High, and that Faubus came over to the school and I killed him with a machine gun . . . or I dreamed one day that I was ambushed by the police in Little Rock and they wounded me and I was killing them, and then the army came and they stopped, and one of them told me that the only reason that I lived was because I was white. Niggers die! I can still picture that one.

In his freshman year at college he did well. The school itself was in some turmoil, with various groups advocating various pursuits in civil rights action.

We studied, but what was happening in Nashville or what happened to those kids in New Orleans was more on our minds. . . . We talked about that like I guess you talked about your courses when you were in college. We'd argue about whether you should stay and get through college, or go out and participate. We'd argue about whether you can serve your race better by just getting educated and doing your work, or whether you should get out there and fight for freedom. I always felt you could separate the men from the boys . . . I thought that this was our big moment in history . . . it wouldn't be worth having a profession and a family if they're going to grow up in semi-slavery. I wanted self-respect more than a degree. I felt the others were cowards . . . no, I don't think that now. I realize that not everyone feels the way I do about this. A lot of our people don't know any better. And a lot are just scared. Some of the guys were scared of ruining what they thought was their big chance at college and their parents' one big hope. I thought of that, but I thought about how can you have a chance when a segregated society treats you only a little better than an animal? . . . My father fought in France for this country, almost got killed. Why should he be afraid to vote? Why should he tell me that I'm not as good as some white kid, and why should he sit and "yes" the white man all day, and then come home and booze it up and tell us how rotten they are? Man, it's not worth it . . . I decided that . . . and I joined up on the freedom ride in my spring vacation. . . . Now I was really doing something, and I couldn't go back. I thought this was the best education and the best service I could perform . . . I'll go back to school in a while, maybe a few more months . . . I just want to make sure that I've really given something to this cause . . . and there are kicks, too. You meet a lot of interesting people. You get to know the reporters. You see yourself in the papers, and really feel that you're almost single-handed in breaking some of those things . . . I get worried and nervous a lot. Mostly tired though, it's tiring, traveling and talking to other students. . . . No, I've never felt depressed. I'm in pretty good shape mentally. I get headaches sometimes, bad ones. Maybe I've got migraine . . . twenty-four times in jail, more than my age!

We do not really know John after we read this. We have some idea about some of the things that have happened to him in his short and eventful life, and we have some idea about some of his feelings. A psychiatrist would talk with him a number of hours and would tell us that he is not deluded, and not distant, and can be understood when he talks. He would tell us that he is intelligent and he would find out that before this young man became involved in his present way of life he had never entered a courtroom, let alone a jailhouse. Hearing intently, he would begin to see how this young man grew up, how he got along with his mother and his father, and, of course, in this instance, his grandmother. He would consider his behavior as well as his words. He would find him to be a tall and thin man who looks a bit younger than his twenty years; who is neat and orderly in appearance and walks slowly and, at times, hesitantly; who talks easily and with warmth; who has enjoyed good health and who has never had cause to feel unhappy enough with himself to see a doctor or a minister. His interests and hobbies and attitudes might be collected, and the roots of some of these in his earlier life might be uncovered. For example:

I think when I go back to school I'll take law . . . it's the best way to fight them [segregationists] when you get older and have to settle down . . . it's the same thing, we're doing one part of the job and they're doing another . . . we need Negro lawyers in the South . . . I'm going to stay right down here.

These comments illustrate, again, the deep involvement of this man's life in his people's struggle for freedom. What is also shown is how his family life engages with the larger, racial issue. He is a strong, tough fighter, like his father, and now in a kind of war, and he wants to go to school and be a lawyer, with dignity and self-respect, like his mother's employer. His personality and his problems are not unique. They are the problems of growing up and finding himself as a man and a working man. What is unique for him in our country is that his skin has grimly attached itself to almost all these problems, to his early years as a child and his later years as an adolescent, to his dreams and actions and fears and aspirations. Trouble with his father, longing for his mother, attempts at achieving independence—all of these are touched by his light brown color. Decisions and choices have to pass the muster of race. If he becomes like his father in many respects, his father's opinions and feelings about the Negro and the white will be there to confront him, as well as how those feelings affected his childhood. And so with his mother, and with his grandmother. Like all people, he will sift and sort and take from many people, within and without the home, those many ways and traits and habits which make up the daily lives of a person. Race, religion, ancestry, affect us all. Saying that the Negro has a special problem in this regard, saying that he is pervasively affected, intensely affected, is merely describing his lot today among us. His lot, in each person, becomes the individual's private burden and challenge. It is hard to see how other burdens and challenges in the individual will escape the magnetism of this one.

But the individual life still exists in Negroes. Each one of these students took his own road, made his detour for private as well as public reasons. It *is* a detour—away from school, into jail, toward danger and possible violence. Many youths, Negro and white, in both North and South, favor what these students actively proclaim, but for reasons in *their* lives cannot or will not join them. Here we are up against history itself, and there are no easy separations or distinctions between these two groups of people—certainly no psychiatric ones. Those who protest are not psychotic, retarded, delinquent. They are all very specific in their protests. Indeed, their college teachers have told me that they wish at moments that a portion of their defiant spirit could be applied to some of the crusty and stunted areas of college and classroom tradition. As you talk with one after another and hour by hour eliminate the broad categories of the crazed, eccentric, lawless, or mentally inadequate, you soon find yourself meeting yourself and your fellow human, that vast body of mankind which is alike and unalike, sometimes sad, sometimes joyful; gifted, and tiresome.

Driven by their own motives and past lives, but united by the historical moment that selects the doers and generates deeds. There is the very bright son of a Negro law professor who could not study well at college. Every college dean knows this problem, the famous father whose son for various reasons cannot assert himself academically. It was in the midst of this kind of difficulty that the students obtained this new recruit, able, forceful, very valuable. After over a year of doing things which very conservatively might be considered brave, he could return with high efficiency and performance to a less overwhelming and frightening college career. Another shy and slow talking boy, an only child, participated in one sit-in, for which he was arrested, lost his job, and with that, the means of support for an aged and sick parent, whose death occurred shortly thereafter. In his fierce grief he cried his accumulated rage of years after his mother's last penniless minutes in the grossly overcrowded, understaffed slum of a Negro emergency ward in the city hospital, by leaving college for a time and walking the cities in silent outrage. These two are unlike in their homes and in their experiences in roughly the same two decades of living. One is bright, well spoken, rebellious, casual in attire and, among Negroes, born to a manor. The other is poor, cautious, very neat and careful of person, and a bit sad and heavy, and unable to initiate ideas or actions. But he will follow more than the instructions of others as he walks the streets.

Walking with the Negro youth are their white friends. It is well known in the North that many white students in our universities have strong sentiments on the subject of civil rights. In the South these Northern students are considered no different than the many others who have hurled accusations in that direction, some unjust, some just, but rather foolishly and inappropriately self-righteous, and all from outsiders. But in the South there are white students who are deeply of the South and who can be seen beside their colored brothers. They can be heard talking about their lives and about their rather special work and special trials. A tall, blond, twenty-two-year-old native Alabaman speaks in a soft and leisurely fashion:

I never thought much about this one way or the other until I was a sophomore at college in Montgomery. I was taking a course in sociology and one of the topics was about the Negro in Alabama. I started doing some reading, and then talked with a few teachers. Many of them would tell us their private opinions about the issue, but were afraid to speak out publicly. I started getting more and more interested. At first it was just intellectual, I wanted to write a good paper and wanted to know about what I was writing about. . . . I started noticing how Negroes existed. It suddenly occurred to me that all the things I took for granted were forbidden to them. . . . Now I was a junior and a big deal on campus, and I talked with more of the professors about this. . . . They seemed scared when I said that if everyone in Alabama said what they really felt we wouldn't have some of these things. . . . I'm not even sure about that, but I do know what I feel, and I have to live with myself. . . . My father once belonged to the Klan, but it had nothing to do with Negroes. I think the whole town joined after the first world war. He's a minister . . . has always said that this whole segregation thing is wrong. He's glad that I'm doing what I'm doing . . . says that if he were my age he'd be out there fighting, too. My mother worries, but she always worries about us. . . . We've been in this state as long as anyone, generations . . . no, it was gradual involvement. I don't remember the exact moment when I went from studying the problem to doing something about it. . . . I've always liked action. I think that if you believe something you can sacrifice for it . . . this is our biggest test here in the South, and we've got to solve it ourselves by realizing that we're not free when we can't say what we want or associate with anyone we wish. . . . I like tennis, swimming, anything that really gives me a work-out . . . I'd like to put in another year at this, and then maybe go back to school and study about all this, like in sociology or psychology. Figure out why people are so strange about all this . . . let their kids be brought up by Negroes, let them serve them and prepare their food, then blow sky high if one of them tries to sit near them in a Walgreen's. . . . We have our problems as whites in the movement. I no longer stay in white hotels . . . I decided that as long as they were segregated and I couldn't be in a place with my own friends, I'd stay in the colored places. I go to the colleges and talk with students. You'd be surprised at how many people are with us, all over the South. They have to be quiet, and do things underground style . . . I never push them, never try to get them to do something that will get them expelled just to prove a point. They can talk in the cafeteria or in the dorms, and they can find out how others feel, and they can work from within . . . very often with the help of a lot of the faculty, who have to keep quiet about it, too. . . . That's what bothers me more than anything else, not being free to speak out and say what I want to. I keep telling some of the fellows I meet that I joined this because if I can't say what I want, I'm not any more a free person than the Negro. . . . I used to debate, was on the team . . . used to think I'd be a politician, run for the Senate, but not much chance of that. . . . I've changed, learned so much about what's really going on around me. . . . They really get enraged at me when they can't just dismiss me as a damyankee. . . . I've been arrested so many times I've given up counting, dozens . . . everything from disorderly behavior, violating city ordinances, unruly assembly, criminal anarchy, the whole range. . . . The jails vary. Some of those police know the handwriting on the wall, laugh about the whole thing with us. Others are real sadists. Turn on the hot air in summer to burn you up, or let you freeze in winter, or mess your food up, and swear, can they swear! Sometimes I think I'm becoming anti-white, I get so disgusted . . . I can understand how some of the guys just get fed to the teeth with the whole white race . . . then we settle down . . . they remember me and a lot of others and we try to forget the whole skin thing between us. I'm black, they're white, we joke about it. I do get discouraged sometimes . . . but I think about what I'm doing, and it really is more significant to me than reading books . . . I can go back to that, but then I won't just feel I've been in an ivory tower all my life. This way I've done something to make this a better country . . . I'm no complete idealist . . . I know you've got to be strategic . . . but there's a time when regardless of what you say to yourself and read, it rings hollow compared with what people have to suffer in the world, and that's the time to commit yourself to action. . . . It took me a few years, but I think you need time to find out about some of these things before you can do anything about them. It took me a long time to figure out what I could do even when I wanted to do something. . . . Sometimes some of my Negro friends think I'm nuts for putting my life on the line . . . I tell them that they've been brainwashed by the segregationists into believing they're not worth the effort, and, besides, this is my fight, because the white race is really doomed if we don't solve this problem.

There is little need for comment here. This boy grew up in a deeply religious home, an active, vigorous, alert youngster who runs well, plays good tennis, reaches for prizes in expressing himself in argument in high school, dreams of political glory, does well in college, is slowly struck by certain underprivileged people, and, in his father's own tradition, stops a while in his life to apply his restless, seeking energy to their plight, which becomes his. He outdistances his father, as each generation may, and he lives with people the life of someone who gives of himself to those less fortunate. They love and, in a sense, adore him, as people do when they meet someone who extends, even if slightly, the possibilities in man.

They also love a young girl whose eyes are hazel and hair long, brown, and flung about when she turns her head in emphasis. Also Southern, she has not left college. But she works as hard there for what she believes as she studies what is taught. A willful girl, she had a less gradual introduction to the unreason of race.

There are a few Negro girls at my college, and, honestly, at first they scared me. I didn't know them, but I just wasn't accustomed to seeing Negroes in a school or college with whites. I was brought up with them all over the house, but that's where they were supposed to be. My mother was always very good to them . . . told us they were like wonderful children with grown bodies that did all kinds of work for us, but not much mind. I believed her for a long time, but I began to question everything when she didn't want me to go to the very college she'd been hoping I'd go to all my life just because they desegregated and took in a few Negroes. You'd have thought the South was collapsing to listen to her. But I went, and one of the last things she told me was to be careful about them . . . kind of like there was a pestilence that might take every white body near it. Daddy just laughed and laughed, and used to tease mother, and tell her she was just plain scared, and letting it all out on the poor nigras. I can hear him saying "Mother, you just leave them alone, and they'll leave you alone. They've done plenty for us, and if their day is coming, it'll be good for us as well as them. Take your fears out on something else besides them. We've used them too long already for that kind of hating and beating. I'd rather kick a tree or something when I'm not feeling good. . . ." Daddy is a lawyer . . . comes from South Carolina, but has always been liberal on race . . . he told me to take anthropology in college and find out about as many people as I could, because they're all going to be heard from in my lifetime. . . . But I was scared, kind of a physical feeling that told me to stay away from them, it's hard to explain. One day one of them was brought to the table where I was eating. I had to get a strong hold on myself to stay there and to keep on eating. I couldn't help it, I just felt sick to my stomach, sick all over. I smiled and tried to be friendly, but I was in a cold sweat . . . later, in my room, I realized that I *was* sick, that it was a kind of sickness when you react that way to another person you don't even know, just because of color . . . it's more than that, though, it's the way you've grown up, and been taught. I said that night to myself that if I could throw away some other ideas I'd heard at home, I could throw this one away too. I just *had* to . . . it was like curing myself. . . . I forced myself at first to get nearer to them, to talk with them or sit with them at the table. Then I found that I really liked one of them, she and I had lots to talk about, and she was a really attractive person . . . then, my old reactions seemed so strange and I can't understand how I ever could have felt that way . . . of course, when I went home and told my mother about this, she almost died. She cried and we really had a time. She swears that she'll send my brother only to a segregated college, but I laugh and tell her that by the time he's ready to go to college she's going to have a tough time living up to her word. Bobby is fourteen, A lot is going to happen in four or five years. . . . Daddy just laughs, and tells me that Mother makes a lot of fuss, but is tough underneath and she thinks she *has* to get excited about things like this. "Your mother hasn't it in her to hurt anyone, black or white . . . she's just changing her ideas in her own way. She has to cry them out, and repeat them a few times so that she can say goodbye to them. That's the South, holding to its own as long as it can, because it had precious little to hold to for a long time. . . ." That's Daddy for you . . . after a while I decided I'd actually *do* something about all of this, and so I started speaking up whenever the subject came up, and you'd be surprised at how many people just are waiting for others to take the lead . . . and then we formed our campus organization and picked certain jobs to do, like education at our own school and helping put pressure on the movie owners and restaurant people. Most of their business comes from students, so we should have some say in how they treat people . . .

She and her friends were having much to say, and their words were expressed in deeds, which caused changes in the admissions policy of local movie houses and cafeterias. In her Southern state this is possible. There are others like her in the bottom tier of states along the Gulf of Mexico who, if they are to stay in college, must be more circumspect, and less hopeful of immediate translation of ideals into realities. If we play back the hours of tapes which have recorded her voice and her words and her story, we can sense certain themes: of her mother and her father, and how they differ and how she gets along with them and expresses her divided loyalties in many ways, not the least of which is her position on the racial question; of her childhood and training and memories as they meet a world which no longer sustains their value or truth or reality; of the crisis in her life, the discord when these old traditions in her meet upon new ways in the South, in her college; of her solutions, accompanied by fear and guilt, and brought home for their resolution; and of her emergence as a grown lady. In this case there is a more decisive experience at the supper table, which engages with not too dormant personal struggles, in themselves universal for college girls. A crisis in history is a crisis for people, and human beings yearn for ways of expressing themselves. This girl found her way.

Since the way of these students involves obvious danger, those taking the way must be able to live with this danger, must be able to persist undeterred by it. Facing danger at their own behest, devoting themselves to its encounter with such fullness and passion, they offer instructive examples of how young people manage such commitments and do not falter. The obvious dangers are from crowds

205

and from the police and from judges and jailers. But there are other problems, too. There are problems of money, needed to eat and sleep and travel; needed to get out of jail while the judgment and sentences are appealed. Bail can be set in the thousands by a local court which so desires. There are problems of the devil: of publicity, of attracting attention to their actions, so that people will be shamed, embarrassed, or forced to act out of expediency, national interest, economic or business anxieties. In a country of interests which often conflict and whose genius is their reconciliation, they too must become versed in the practical and possible. But this means practical and possible for them, and if they listen too closely to the wise counsel of even their most friendly elders, they might well lose one of their greatest possessions—their own, youthful, sometimes heedless and blind momentum. The problems of momentum are problems of the flesh: of how to behave before mobs, of how to behave with one another, of how to live with oneself, and keep one's morale high.

A psychiatrist interested in how people manage these tough assignments gets to know the young men and women and watches them in action. In this age of machines his impressions can be stimulated and recalled in later months by tape recordings, and he can hover over the tapes the way others may pore over books, looking for some of the thoughts and feelings which may pass him during the conversations themselves. After he has talked to enough of the students, long enough, he gets some ideas about how they manage. How they manage can be discovered without tape, but with tapes we can hear their every word; such as when they respond to questions like "How do you feel when you walk into a situation that you know in advance will lead to violence or arrest or jail?"

I don't feel, I just go ahead. . . . We sing and encourage one another. . . . We've practiced and rehearsed. It's like I suppose my dad did in France against the Nazis, you're in a kind of an army, and you've got no choice, except that you've joined on your own, not been drafted.

I close my eyes mentally. . . . Sometimes I pray just before, or I keep on saying that nothing too bad can happen in America. . . . We have the reporters nearby and the federal government . . . they can't do what they used to do . . . you know . . . lynchings are harder for them . . . I know some of them would like to, but they're afraid to . . . not in this world, they can't . . .

The young wife of one of the students insists:

I'm prepared to give up my life . . . I've made that decision, and it's almost as if I'm no longer in my body. The body can be sacrificed. I just look at myself and say to myself "Man, you can lose your body, but they can't take your soul away. They can just lose theirs. And I'm going to help them. . . ."

Another college sophomore, a young girl who wants to be a teacher someday but has taken a few months' leave hesitates, then starts with:

Every time I feel afraid I just remind myself that we've got nothing to lose. I think of all the things I've had to put up with in my life. I think of all the movies I can't go to, all the restaurants I can't eat in, all those separate rest rooms and water fountains, and, I tell you, I can get so angry that they could have atom bombs on their clubs and police dogs and I'd keep on walking or sitting. We've got our rights as human beings to gain, and absolutely nothing to lose, not a thing. So, why not? Oh, I have to get myself to thinking about this sometimes, when I get nervous, or when the jail is a bad one, and you don't know what they're going to do . . . you can do that, though, get yourself primed, like a pump or something . . .

A veteran of eighteen arrests commented:

It's almost like going into an army. Well, a better example is the Crusades, where men went off for their religious convictions, almost voluntarily. . . . You not only feel what you've always felt, but you've decided to stake your life on the line for it. You'll go to jail, face the police and those screaming seggies . . . face the whole world. . . . We help one another, keep our strength up with music and we talk a lot *after* it's happened, and joke and get some laughs out of it . . . it's fun to see your picture in the papers, and read the different accounts, a lot of them get all botched up, or completely slanted, of course . . . one thing I've heard some of my friends say is that they never really believe that those crowds will hurt them . . . heckle and scream, yes, but not really hurt. I can't quite pull that off in my mind. I know that they may hurt me, but I figure we can take it, and give it back. I've taken worse all my life . . . this is the first chance I've had to *do* something while taking it that may end the whole rotten mess . . .

Clearly, their feelings and their consciences impel them. Clearly, also, in order to endure inevitable apprehensions, they must either remove themselves or temporarily blot out some real threats. Determined minds can do this. They can abstract the mind and emotions from the body, and let *it* serve duty and risk injury. They can deny danger, wipe out or minimize hazards. They can make thorny dilemmas very simple, and they can undertake such activity, such continual movement and travel, that there is little time for nervous brooding. They can encourage singing, meeting, and praying with others, gaining the well-known strength of camaraderie. They can see fear and terror in others, in the enemy, and fail to see anger and rage in them. The enemy becomes afraid, and they are unafraid. They can urge talking and sharing in words the common dread. They can gain very human pleasure from anticipated rewards of reknown, attention, approval, balancing with these all the present hesitations and doubts. They can fall back on what they call "techniques," dwelling upon the details of how to behave, where to go, what to say or do, placing their feelings in an envelope of rituals and memorized, almost automatic, maneuvers. They dribble away large doubts in small annoyances. Panic can splinter into a generally fretful day. Angry jokes can be told. Books which give comfort or release or escape can be read. There is always good food, and one can sleep longer hours. A bad headache, some vague stomach pains, these can silently express unallowed anxiety. There are daydreams, fantasies, nightmares, all perhaps lowering some pressures which may build up. Fun, movies, exercise, may help; and there can be nervously short and intense romances, almost as in war, when the lovers know they may be separated, that they have only a brief time, to which they bring so many emotions besides love. Letter writing helps. And some of these minds can not only feel frustrated and anxious and angry and fearful, but they can hate. Often, if the alternative is despair, they can hate and hate mean and hard, can give back what they've received rather than sink into it themselves.

In courts and in jails their minds have additional difficulties. They are being accused and, usually, condemned. The charges range from disturbing the peace and loitering to criminal anarchy. Those under sixteen or eighteen may be called delinquents and sent for long periods of so-called evaluation and confinement. Some courts have even referred them to mental hospitals for psychiatric observation. With such certain punishment for their efforts, they must learn how to live in cells and under charges of criminality or insanity without losing their sense of themselves and their integrity. This requires no little effort. To do this they can deny their guilt, and turn the illegal into the just. You can hear them listing with pride their arrests and jailings. You can hear them upholding their own values against those of cities and states. You can hear them calling for the support of their government and the tradition of Western democracy. They do this out of conviction and out of need. All the time they are being told by people in black robes and people in uniforms carrying guns, by newspapers and by influential citizens, that they are bad and unruly and outcasts. All the time they are being hustled and confined, after being refused and insulted. This cannot fail to affect them, cannot fail in some part of their minds to make them feel worthless or wrong or guilty as charged. Since all of us have gone through childhood and learned about being good and being bad, and doing good and doing bad, and being praised and being criticized, memories and anxieties of the past can return under such evocative and haunting circumstances of the present. In a sense they are being sent to a room, which is now a jail, to be alone and to repent for being bad and disobedient. This is precisely what one of the students dreamed while in prison.

Living the life of the condemned asks something of you. It asks you to have control, just as non-violent protest requires control, except that now you are not free any more. Control of feelings and behavior is important to these students, and in order to insure its strength they often live very ascetic lives, as if the large amount of control needed must be obtained by spreading it into almost all of their activities. In some of the front lines of Mississippi, relaxation of body or mind must be guarded. Even when they are not protesting they may be under constant surveillance. They must drive cautiously and live cautiously. Even among Negroes they must keep an image of dedication, because suppressed people are suspicious and afraid. They must control the temptation to exploit their work. Once they become heroes to some of their people easy promiscuity and many comforts may be offered them. They are anxious, and part of them responds very humanly to it. One can hear them in meetings talk about this, warn themselves in advance and point out the dangers:

If we're going to stay in the Movement, we've got to watch ourselves, we've got to control ourselves all the time. It's like a war. They call it a fight to the finish, and so will we. If we're going into Mississippi and Alabama, then we've got to have iron discipline. . . . You either put all your energy into the Movement, or you stay behind and so do something else . . . no fooling around, we can't take risks like that, we've got enough with the "segs" without setting things up for them ourselves.

At times, under extreme danger, they go into a state which they call "commandati," which means extreme alert and caution, which measures a response to all the harassments that accumulate in a struggle such as theirs. It is not that they are by nature unruly or wild. On the contrary, they tend to be rather controlled and studious people. It is that they are under attack, and they must prepare themselves and one another for such unusual strains. To hold up under these strains requires devotion, energy, and ingenuity. Perhaps only the young, the still unattached and unemployed and unsettled, can take this on.

They take on cells for homes. They must hold their sanity, keep their orientation, amidst isolation and obvious humiliation. It is one thing to go to jail for a crime committed. Many who do can even shrug off the jailing as part of the game, part of the gamble. It is another thing to have a highly developed conscience, and to be moved by that conscience to certain deeds of fulfillment, and in pursuit of these, in quest of being a fuller citizen, to be insulted, fined, sent off to solitary confinement or overcrowded cells. It is another thing because these students are not hard and tough and callous, and yet they must get along where they are often treated with more contempt than hardened lawbreakers who are there for robbery, assault, or drunk and disorderly behavior.

In a Louisiana prison I talked at length with a tall, thin, twenty-year-old student who had been held in solitary confinement for two months on a charge of criminal anarchy. When all the doors had been closed behind me, the first view was of a large blackboard with the prison census divided by races first, sexes second. I wondered what genes or chromosomes made for such high numbers of male Negroes and low numbers of female whites. I was there at the request of his counsel, because this young man had written letters to friends in Atlanta requesting psychiatric help saying, "I'm beginning to think I may be cracking, I'm really getting near the edge." His criminal and anarchic behavior was organizing students in a Negro college nearby, urging them to help other Negroes vote, or be served in department store lunch counters. When he first came to the jail he was put in close quarters with recurrent criminals, the most beaten, desperate, and lowly of his people. In the medical examining room, and later after his release, he talked about it all:

You know, I thought until then I knew how to get along in jail. I'd been in lots of jails . . . but this was almost too much. First I had to fight to keep away from those guys . . . sodomy all the time, and I think they were egged on by the guards. What a bunch of thugs. They told me that the place had been investigated twice before, so they were going to be careful, but they let me have as much as they could . . . sent heat in the cell when I was hot anyway . . . finally moved me to solitary. I thought at first that would be better, but its gets strange after a while. You fight just to keep track of time and to get *through* time. It drags, man, it drags 'till you think you've just about had it. You try to write, if they'll give

210

you paper. You dream, and try to sleep. But I couldn't sleep. No exercise, so I didn't need sleep. If only I could have slept, or dozed all day. . . . Terrible slop for food . . . I'll tell you what it does to you, it makes you wonder whether you'll ever get out, and soon you begin to lose some of your fight. They would get their trusties to come and shine their shoes in front of me. Those pitiful Negroes, on their knees shining the shoes of the guards, and then they'd ask them, "Joe, aren't you happy with things the way they are in Louisiana?" And big Joe, big, free, independent, American citizen Joe, would look up at them and say, "Yesuh, yesuh, Mr. Boss, I sure am, yesuh." Then they'd turn and make the poor guy tell that to me. . . . The worst was when they brought those kids in, those white school children and pointed me out as a Communist. They had me on display, like a damn guinea pig . . . I was marking the days off on the wall . . . they finally gave me two books . . . Kipling's short stories and a history of China written in 1849 or so . . . yeah, pretty funny, but not then it wasn't. . . . Those pills did help, they took the nerves out of me, let me settle down . . . they told me they didn't *have* to give them to me. . . . The doctor came around and asked me why I didn't tell him about what was aching me or bothering me . . . I couldn't trust him . . . he was probably O.K. . . . made sure I got the pills. I'm glad they let me write to you. Writing a letter can keep you going for days . . . I don't know if I can take another spell like that. I'm not crazy, but I began to wonder how many days I had left before I started getting there. . . .

He was indeed very anxious, and more frightened than he may have known. He was afraid of assault—sexual, personal, moral, mental, and spiritual. He was afraid of losing the complicated organization of mind and emotions which we spend so long building and which is not meant to be tested by hells like this. He was afraid of losing his gritty, tough, daring ways and dissolving in tears of panic and confusion, if not disorientation and delusion. When you are alone for a long time, and you know that people want to hurt you and have that ability in their absolute power, every sound, every shadow, every movement, can signal danger and death. Fortunately he was released after three months in jail. I think it was good for him to be seen by a doctor and put on tranquilizers and allowed to write. His jailers and their superiors in the nearby district attorney's office could scarcely let him effectively lose his mind. That might hurt *their* cause. Such was his protection, and such is the way men are protected from men in some places.

Sometimes they spend short periods of a day or two in jail. Often they can turn these days into frolic. They are together in cells, and they sing and laugh and talk. Or, if alone, they can send messages and whistle and sing back and forth. Many flippant, arrogant, angry phrases, sarcastic phrases, words of mock irony can be heard; sometimes smouldering resentment, neatly kept, can become a white heat of rage and hatred. This can cause shame and alarm for those who are not Negro, and are torn between their sympathy and their ability, because of different experiences, to know better, feel less desperate.

It's hard for the white student in this at times. Sometimes you have to keep quiet and listen, let them get it all out of their systems every once in a while. I feel terrible . . . it's no good if we get desegregation and they're so embittered they become just like the segregationists. That's the danger, now. Look at the Black Muslims. Some of the guys I know are not Black Muslims, but they're becoming kind of African Nationalists . . . they tear down this country and say that Africa's the place for them, to live and . . . where everything is good. That's what happens when you treat people like this . . . most of them really love this country though . . . I don't know if I would . . . after a while they cool down . . . sometimes apologize to us. . . . I understand how they have to get like that sometimes . . . I don't take it personally. I think the Movement should have whites in it, even in Mississippi, where they get so enraged when they see us together . . . that's what it's all about, that we can work side by side.

They work side by side, and then a time comes when they leave, going back to college, to work, or to a graduate school. When they decide to leave, like when they decided to come, will vary from person to person, as will the length of their stay. Some have been working at the problem for years, some for months. They all recognize that one cannot spend a lifetime at this, at least not in the capacity of a student. Perhaps more of a problem now for them is the inevitable consequence of their various successes. Like any gathering of human beings who are trying to do things together, they have to come to some settlement with the nature of themselves as a group or organization. In 1960 when four of them went into a store in Greensboro that was not a problem. It still isn't in the many spontaneous instances where a group of high school or college students simply decide to take on a specific job, like a movie house in their town. But if a systematic assault is to be made through direct protest, and if considerable resistance is met, requiring planning and more time and effort than originally estimated, there must be distribution of work and energy and direction. Living in a highly organized society, the students learn that it will not necessarily fall under an attack of the moment. And so they pool efforts, and since there are many colleges, many cities, and many states, they come together and meet and talk and argue, and try to regulate themselves, and they belong to different organizations and call themselves by different names.

212

They find, in the continuing irony of human existence, that adversity has its own joys and success its own trials. They worry about losing their freedom and spontaneity in a mass of constitutions and by-laws and regulations. They worry about exhausting themselves in bureaucratic tangles, in endless appeals for money. They fear the futile absorption in stamp machines, mimeograph machines, and filing cabinets, even as they know how much these are needed. They struggle with these new and petty tyrannies, trying for the needed balance of personal action and staff action. This struggle may be harder, in the very long run, than even that for the rights of the Negro. Wherever there are men there are rivalries, envies, competing needs for power. Wherever there are men there are also possibilities that great and decisive things will be done and ideas conceived. They are struggling for the right to be human, to be men in the many senses of that word, and so this, too, will mark their progress; when they can wrangle with the fruits of their labor like all other free men.

Their harvest becomes a harvest for all and a source of challenge to many. Those who are interested in how people do things to change things around them will study these students. The various social scientists will consider their case. Certainly they offer interesting challenges to psychiatrists. Psychiatrists are asked today to help the courts decide, not only whether a person knows the difference between right and wrong, but also whether he is so constituted that he can adhere to this understood difference. We have urged considering motives and impulses as well as facts and strict rationality. We have urged changes in laws where we have felt them to be unfair or harsh or not in accordance with our view of the nature of the human mind and its development. We urge understanding of people, and help for people who are not well. Finally, we are asked, not only to do our healing and our research and teaching, but to advise the general public, as well as the courts, on the widest variety of subjects. We are consulted about the nature of delinquency, what it is, who the delinquents are, and what can be done. We venture more and more into prisons to work with prisoners, on the good premise that people impelled to crime have some disorder of impulse as well as a record of wrongdoing, and should be treated with more than custodial reckoning. We are asked about problems in schools and colleges, problems of learning, of behavior, of suitability for certain programs. We are consulted by many professions, to help them choose candidates for the ministry or to screen applicants for crucial places in business or the government. Presumably, then, our capacity to evaluate human beings and their behavior is considered by others and by ourselves to be of some practical as well as theoretical value. Presumably, also, our knowledge is felt to extend beyond the mental hospitals and consultation rooms and into the life of the community.

The lives of many of our communities today are endangered by troubles between groups of people. Large numbers of people gather themselves around certain names describing religions or nations or sections of nations or races, or combinations of these. They declare themselves apart by rituals or flags or color or customs or locations. In their behavior with one another these multitudes are often like individuals: they get along or they fight, they help one another or they try to control one another. It has been this way for a long time. New collections of people emerge and old ones die as new bonds and ties appear and old sources of allegiance seem less compelling. It has been and still is a part of growing up in most of the world to find out about oneself, and many of these associations—national, sectional, racial—help in this and contribute to it through traditions, customs, knowledge, security. Perhaps when people know themselves, and are not too afraid of themselves, and are relaxed and at ease, they do not have to lose themselves in overworked assertions of themselves as members of particular groups rather than as persons.

Certainly these students have shown that American youth can look around and see things that are hurtful to themselves and to others, and venture forth to change them. Certainly their work is important, and certainly it should be evaluated if we are concerned with changing life in our community. As psychiatrists we have this concern, and our society asks this concern of us. Our challenge with these students is, as always, to gather observations and descriptions; then attempt to sort them out and come to some estimate, always tentative, of what their behavior means and what it may portend. To do his may be easier than to reconcile some of our own professional problems. But, in a stimulating fashion, the behavior of these students may help clarify some of our professional problems.

Not the least of our problems is that elusive idea, "normal." We wonder what it is and how to use it; we wonder whether it exists and we are struck by the curious obsession which it inspires in many around us. Not the least of our problems is deciding what is "sick" and what is "healthy." We should use these words carefully, too.

Then, we have to decide at times who can best take on a particular job or perform well in a certain situation. We thus start evaluating "creative" and "destructive" parts of the person, and find them often entangled enough to blur their separate meanings. We also have to worry about our own values, our own hopes and ideas of what is desirable for man, what helps him grow, and what is crippling and harmful.

How, then, do we comprehend these protesting students who break laws and go to jail and are sent to detention centers as delinquents or to mental hospitals for suspected insanity? How do we evaluate their departure from college, their radical departure from established customs and habits, their defiance of law and order, their public display of themselves and its resultant public uproar? How do we evaluate their desire to do things even though they know they will be insulted, attacked, injured? How do we evaluate this stubborn, sly, systematic assault upon the laws and conventions of our society by youths of both races? Casting a glance elsewhere, we can ask some other questions. How do we comprehend those students who don't participate in these demonstrations? How do we evaluate their attitudes toward themselves if they are Negro, or toward others if they are white? How do we evaluate their willingness to endure, or see endured by others, these restrictions and deprivations of person, property, and dignity? How do we look at them in school, or in their obedience to laws which curb and isolate them or others? Finally, do we dare look at our society and ourselves in it with any questions? What in laws and customs will help people to be secure or to maintain health, and what is injurious? What do we mean when we use the words "well adjusted" or "poorly adjusted"? We certainly use them. Should a boy who has been arrested twenty-five times and put in jail eighteen times and charged by his society with criminal anarchy be called "well adjusted," "delinquent," "anti-social," or "creative"? Does he have a "problem with authority," and, if he does, why don't more people have it? Should we call it a problem? A problem for whom? Students like these challenge the unrestrained application of adjectives. Such nouns as "masochism" and "acting-out" are also challenged by these students. Are they "acting-out," are they masochistic, are they troublesome deviants? Or are they other words we use, like well integrated, or "mature," with "good defenses" and highly developed sublimations? Finally, what do we do about applying what we know about people—what is healthy and sickly in them—to these problems? These students are

215

denounced and praised. Should we, can we, evaluate their actions with some sense and fairness? Or do we find such problems too prickly, too dangerous to our positions in society, too risky for involvement? Perhaps these students challenge our assumptions, our concepts, our language, our present ability to answer many problems; they make us aware of some of our limitations, despite society's importunate requests of us and some of our indiscreet replies. If so, they will have given us a measure of humility. This would be a large gift. With humility, we need not apologize for some of our present ignorance. Ignorance is a challenge, and it is a high honor to engage with it. We lose honor only when we see victory when there is still a long struggle ahead. Perhaps these students of social change and we students of human nature can meet in that struggle.

For there can be no doubt that many of their problems are also ours. Daily they show us how they must refuse the arbitrary and absurd confinements of skin color. Each of them is an individual, and they hope that someday they will be known to their fellows by themselves, by who they are and what they do. Pigment seems to them, to those who are white as well as black, a frivolous and tragic standard for human knowledge of one another, an irrelevant distinction, indeed. It is, then, the meaningless word, the hollow hurtful evocation which bothers them. With us, too, categories and labels of soothing certainty abound. We have no little ability or wish to add to these terms and to these conceptual separations of the mind within a person and of people from one another. Our peril is that, driven by the need to find out about ourselves, we may find ourselves more puzzled than ever by our own thinking and the conflicting faiths to which our thoughts adhere. We must challenge our concepts and categories just as the non-violent youth of our South challenge those of others. If our terms fail to describe what is happening in the world of people and their lives, they must be discarded. If they are contradictory, inadequate, misleading, they must go. If they fail to account for human effort and courage and man's deeds toward a more decent world, if they fail to envision and honor the reality of the heroic in the smallest, quietest assertion of a deeply ethical youth at a Southern lunch counter or in a Southern jail, they must depart. If these students will help to draw us to such considerations and clarifications, their efforts will be our accidental but most fortunate and timely gain.

ROBERT JAY LIFTON

Youth and History

Individual Change in Postwar Japan

YOUTH CONFRONTS US with the simple truth, too often ignored by psychologists and historians alike, that every individual life is bound up with the whole of human history. Whether or not young people talk about their historical involvements—Americans usually do not, while Japanese tend to dwell upon them—these involvements are inevitably intense. For those in their late teens and early twenties find themselves entering, sometimes with the explosive enthusiasm of the new arrival, into the realm of historical ideas. And they bring to this realm their special urge toward development and change.

In Japan, the rather sudden emergence of outspoken "youth attitudes" has led to facile generalizations about the nature of young people's contemporary historical experience. There is first the claim (perhaps most popular in the West) that nothing is really changing, that although things may look different on the surface, deep down everything (and everyone) in the "unchanging East" is, and will continue to be, just as it (and they) always have been. And there is the opposite assertion (a favorite of Japanese mass media) that young people have changed absolutely, and beyond recognition, so that they no longer have any relationship to their country's past. To avoid these polarities, I have found it useful to think in terms of the interplay between inertia and flux in cultures and individual people as well as in inorganic matter. For in Japan one discovers that inertia (maintained by traditional psychological patterns) and flux (stimulated by pressures toward change) can both be extremely strong—that individual change is at the same time perpetual and perpetually resisted.

In my work with Japanese students[1]—done mostly through

217

intensive interviews—I have tried to focus on ways in which they experience and express the wider historical change taking place within their society. I have looked for consistent psychological patterns among them and have then tried to understand these patterns as both old and new, both specific and universal. That is, each is related to the psychological and social currents of Japanese cultural tradition; to psychobiological tendencies common to all mankind; and to forces of historical change, particularly modern and contemporary, in Japan and throughout the world. (I shall refer in this article mainly to young men, since they are most directly involved in the historical issues under consideration. The discussion applies to young women too, but the special features of their changing situation require their own full treatment.)

It is impossible, of course, to make an exact determination of just how much the cultural, universal, or historical factor is at play. But I have found it necessary to take all three into account in order to gain perspective on any immediate observation. This form of perspective seems particularly relevant for Japanese youth, but it is perhaps no less relevant for any other age or cultural group. I have also stressed the *direction of change*, on the assumption that the psychological experiments of outstanding young people can to some extent anticipate future directions in which their culture at large will move.

Historical Dislocation

The most fundamental of these patterns is the absence in contemporary Japanese youth of vital and nourishing ties to their own heritage—*a break in their sense of connection*. It is not that Japanese youth have been unaffected by the cultural elements which had formerly served to integrate (at least ideally) Japanese existence— by the Japanese style of harmony and obligation within the group life of family, locality, and nation; and by the special Japanese stress upon aesthetics and the liberating effect of beauty. Indeed, such elements are all too present in the mental life of young Japanese. But they are now felt to be irrelevant, inadequate to the perceived demands of the modern world. Rather than being a source of pride or strength, they often lead to embarrassment and even debilitation.

This lack of a sense of connection extends to their view of the contemporary society which they are preparing to enter. The word

218

"feudalistic" *(hōkenteki)* comes readily to their lips, not only in reference to rural Japan but to "Japanese" forms of human relationship in general; and "monopoly capitalism" *(dokusen shihonshugi)* is the derogatory phrase for the modern—one might almost say postmodern—society that dominates the large cities. Underneath this semi-automatic Marxist terminology is the profound conviction of the young that they can connect nowhere, at least not in a manner they can be inwardly proud of. "Society" is thus envisaged as a gigantic, closed sorting apparatus, within which one must be pressed mechanically into a slot, painfully constrained by old patterns, suffocated by new ones.[2]

Yet the matter is not quite so simple. What is so readily condemned cannot be so summarily dismissed. By turning to two individual examples, we can begin to recognize the inner paradox and ambivalence of this historical dislocation.

A student leader (whom we shall call Sato) in his early twenties described to me the following dream: "A student [political] demonstration is taking place. A long line of students moves rapidly along . . . then at the end of the line there seems to be a festival float *(dashi)* which other students are pulling." Sato laughed uncomfortably as he told his dream, because he could begin to perceive (as he explained later) that it seemed to suggest a relationship between student political demonstrations and traditional shrine festivals. This embarrassed him because such political demonstrations and the student movement which sponsored them (the *Zengakuren*, or All Japan Federation of Student Self-Governing Societies) had been for the past few years the central and most sacred part of his life, in fact the only part that held meaning for him; while a shrine festival, symbolized by the large float, seemed to him something quite frivolous, or worse. He was particularly struck, and dismayed, by the fact that it was *students* who were pulling the float.

In his associations to the dream, he recalled the shrine festivals he had witnessed in the provincial city where he had attended high school; these he remembered as dreary, unanimated, motivated only by commercial considerations, and ultimately degenerate, stimulating in him feelings like those he sometimes experiences when face to face with very old people—a combination of revulsion, sympathy, and a sense of contamination. But he contrasted these negative impressions of relatively recent shrine festivals with the romantic and beautiful atmosphere of great shrine festivals in the distant past, as described in many court novels he had read. And he also thought of smaller festivals held at harvest time in the rural area of central Japan where he was born and had spent his early childhood. He spoke vividly of the sense of total relaxation that came over the entire village, of the bright decorations and gay atmosphere around the shrine, of the exciting horse races made up of local entrants, of big feasts with relatives, of masked dances *(kagura)* giving their renditions of the most ancient of recorded Japanese tales (from the *Kojiki*), of fascinating plays performed sometimes by traveling troupes *(ichiza)* and at times by young people from the village. Sato emphasized that in his dream he was a bystander, standing apart from both the political demonstration and the festival-like activities. This he associated with his recent displacement from a position of leadership within the student movement (because of a factional struggle) and with his feeling that he had failed to live up to his obligations to colleagues and followers in the movement. One meaning he gave to the dream was his belief that the student movement, now in the hands of leaders whom he did not fully respect, might become weak and ineffectual, nothing more than a "festival."

But the dream suggested that Sato was a "bystander" in a more fundamental sense, that he was alienated from those very elements of his personal and cultural past which were at the core of his character structure. These same elements—still the formative essence of his developing self, or self-process—had not only lost their vitality but had become symbols of decay. The dream was partly a longing for childhood innocence and happiness, but it was also an effort at integration. Thus in his nostalgic associations Sato commented that if he really did ever see students pulling a *dashi* in that manner at the end of one of their demonstrations, "I would feel that the world was stabilized," by which he meant in a personal sense that if he could harmoniously blend the old things he carried within himself with the new things to which he aspired, *he* would be stabilized. Like so many young people in Japan, Sato outwardly condemns many of the symbols of his own cultural heritage, yet inwardly he seeks to recover and restore those symbols so that they might once more be "beautiful" and psychologically functional.

Another frequent individual pattern demonstrating the break in the sense of connection is one of exaggerated experimentation, of exposing oneself or being exposed to an extraordinary variety of cultural and ideological influences, each of which engages the young person sufficiently to affect his developing self-process, but never with enough profundity to afford him a consistent source of personal meaning or creative expression. Consider the confusing array of identity fragments (as numbered below) experienced by one rather sophisticated Tokyo-born young man whom we shall call Kondo—all before the age of twenty-five.

As the youngest son in a professional family, he was brought up to be (1) a proper middle-class Japanese boy. But when he was evacuated to the country from the age of eight to eleven during and after the war, his contacts with farmers' and fishermen's sons created in him (2) a lasting attraction to the life and the tastes of the "common man." He was at that time (3) a fiery young patriot who was convinced of the sacredness of Japan's cause, revered her fighting men (especially his oldest brother, a naval pilot saved from a *kamikaze* death only by the war's end), accepted without question the historical myth of the Emperor's divine descent, and "hated the Americans." Japan's surrender came as a great shock and left him (4) temporarily confused in his beliefs, but toward the first American soldier he met he felt curiosity rather than hostility. He soon became (5) an eager young exponent of democracy, caught up in the "democracy boom" which then swept Japan (especially its classrooms) and which seemed to most youngsters to promise "freedom" and moral certainty. At the same time, Kondo also became (6) a devotee of traditional Japanese arts—skillful at singing and reciting old Chinese poems *(shigin)*, passionately fond of old novels, and knowledgeable about *kabuki* drama and flower arrangement *(ikebana)*.

During junior high school and high school years he was (7) an all-round leader, excelling in his studies, prominent in student self-government and in social and athletic activities. Yet he also became (8) an outspoken critic of society at large (on the basis of Marxist ideas current in Japanese intellectual circles) and of fellow students for their narrow focus on preparation for entrance examinations in order to get into the best universities, then get the best jobs, and then lead stultifying, conventional lives. He was (9) an English-speaking student, having concentrated since childhood on learning English, stimulated by his growing interest in America and by the size, wealth, and seemingly relaxed manner of individual Americans he had met and observed. Therefore, when he found himself unaccountably (10) developing what he called a "kind of neurosis" in which he completely lost interest in everything he was doing, he decided to seek a change in mood *(kibun tenkan)* by applying for admission to a program of one year of study at an American high school.

He then became (11) a convert to many aspects of American life, enthusiastic about the warmth and freedom in human relationships, and so moved by the direction and example of his American "father" (a Protestant minister and courageous defender of civil rights during McCarthyite controversies) that he made a sudden, emotional decision to be baptized as a Christian. Having almost "forgotten" about his real family, he returned to Japan reluctantly, and there found himself looked upon as (12) something of an oddity—one friend told him he "smelled like butter" (the conventional Japanese olfactory impression of Westerners), and others criticized him for having become fat and somewhat crude and insensitive to others' feelings. Eager to regain acceptance, he became (13) more aware than ever of his "Japaneseness"—of the pleasures of drinking tea and eating rice crackers *(senbei)* while sitting on floor mats *(tatami)* and sharing with friends a quiet and somewhat melancholic mood *(shoboi)*, particularly in regular meetings of a reading group to which he belonged.

Yet he did not reintegrate himself to Japanese student life quickly enough to organize himself for the desperate all-out struggle to pass the entrance examination for Tokyo University, failing in his first attempt and thereby becoming a (14) *rōnin* (in feudal days, a *samurai* without a master, now a student without a university) for one year, before passing the examination on his second attempt.³ Once admitted to the university, he found little to interest him and rarely attended classes until—through the influence of a Marxist professor and bright fellow-students in an economics seminar—he became (15) an enthusiastic *Zengakuren* activist. His embrace of the *Zengakuren* ideal of "pure communism," to be achieved through world-wide workers' revolutions, and his participation in student demonstrations and planning sessions gave him a sense of comradeship and fulfillment beyond any he had previously known. But when offered a position of leadership during his third year at the university Kondo decided that his character was not suited for "the life of a revolutionary" and that the best path for him was a conventional life of economic and social success within the existing society.

He left the *Zengakuren* and drifted into (16) a life of dissipation, devoting his major energies to heavy drinking, marathon *mahjong* games, and affairs with bar girls. But when the time came, he had no difficulty (because of his Tokyo University background and connections, as well as his ability) in gaining employment with one of Japan's mammoth business organizations. His feelings about embarking upon (17) the life of the *sarariman* (salaried man) were complex. He was relieved to give up his dissipation and find a central focus once more, and in fact expressed an extraordinary identification with the firm. He stressed the benefits it bestowed upon the Japanese economy and the Japanese people, and sought in every way to give himself entirely to the group life demanded of him—to wear the proper clothes, behave appropriately toward superiors and colleagues, and effectively flatter customers (allowing them to seem most popular with bar girls and to win at *mahjong*). At the same time, he retained a significant amount of inner despair and self-contempt, the feeling that capitalism was "evil," and that he himself had become a "machine for capitalism." He had fantasies of total escape from the restraints of his new life, including one of murdering a Japanese or American capitalist, stealing a great deal of money, and then spending the rest of his life wandering about Europe and America amusing himself; and he would also, in unguarded moments, go into tirades against the constricted life-pattern of the "typical salaried man" (*sarariman konjo*).

He attempted to resolve these contradictory feelings by making plans to introduce reforms into his firm that would ultimately encourage greater individual initiative, promote efficiency, and allow for more genuine personal relationships; toward this end he began a study of American writings on human relations in industry. At the same time he was constantly preoccupied with promoting his rise within the firm, with becoming in time a section head, a department head, a member of the board of directors, and if possible not only the president of the firm but also one who would be long remembered in its annals and who would come to exert a profound influence upon all Japanese economic life.

To be sure, neither Sato nor Kondo can be said to be "typical" of Japanese intellectual youth; rather, they express in exaggerated form the experimental possibilities to which all Japanese youth are exposed. (Relatively few become *Zengakuren* activists, but all are confronted with the *Zengakuren* moral and ideological claims which dominate the campuses; even fewer get to America, but none is unaffected by postwar American influences). Even in the majority of youth, who seem to plod unquestioningly through university and occupational careers, there is something of Sato's quest for the past as he works for a revolutionary future, something of Kondo's diffusion, sudden shifts in ideological and group loyalties, and final ambivalent compromise.

What about the family relationships of Japanese youth? Is there a break in connection here as well? I have found that virtually all my research subjects—whether brilliant students, playboys, plodders, or *Zengakuren* leaders—tend to remain very much in the bosom of their families, nourished by the readiness of Japanese parents to cater to their children's wants and encourage dependency—even when such children have reached manhood or womanhood. This continuity in family life seems to be the balancing force that permits Japanese youth to weather their confusing psychological environment as well as they do. But the continuity is only partial. On matters of ideology and general social outlook, most Japanese students feel completely apart from their parents. A typical constellation (actually experienced for a time by Kondo) is the following: the "radical" son remains on intimate (in fact, mutually idealized) terms with his mother; she is sympathetic to his point of view, confident of the "purity" of her son and his fellow students, although understanding little of the intellectual issues involved; his father, with no firm ideological convictions of his own, disapproves, silently, ineffectually, and from a distance, so that father and son are rarely in open combat. The son's emotional state is less one of "rebellion" than of continuous inner search.

What, then, are some of the wider historical factors associated with this break in the sense of connection? We must first look back beyond World War II and the postwar period to the latter half of the nineteenth century, and particularly to the Meiji Restoration of 1868. Before then, Japanese culture, although by no means as even and consistent as sometimes painted, had maintained an effective stress upon lineage, continuity, and on long-standing Japanese and Chinese moral principles—cemented by the extraordinary experience of more than two hundred years of nearly total isolation from the outside world. At the time of the Meiji restoration, however, the Japanese faced a very real danger, not only of being militarily overwhelmed by the West but also of being ideologically, institutionally, and culturally overwhelmed as well. The early slogans —"Revere the Emperor, Repel the Barbarian" (*Sonnō-jōi*); and "Eastern ethics and Western science" (*Tōyō no dōtoku, Seiyō no gakugei*)[4]—and the ensuing pattern of an uncritical embrace of things Western, alternating with recoil from them in fundamentalist horror, revealed *the continuing effort to reassert Japanese cultural identity within a modern idiom.*

Thus, ever since the time just before the Meiji Restoration, Japanese, and especially educated Japanese, have looked to the West with a uniquely intense ambivalence. They have felt impelled to immerse themselves in Western ideas and styles of life in order to be able to feel themselves the equal of Westerners, and at the same time they have waged a constant struggle against being psychologically inundated by these same Western influences. In the process they have experimented with a greater variety of ideas, of belief-systems, of political, religious, social, and scientific ways of thinking and feeling than perhaps any other people in the world. And they have as individuals learned to move quickly and relatively easily from one of these patterns to another, to compartmentalize their beliefs and identifications and thereby maintain effective psychological function.[5] (We would expect an American youngster who actively experienced as wide and conflicting an array of personal influences as Kondo to be incapacitated by his identity diffusion.) Japanese youth are still engaged in the psychological-historical struggle carried over from the time of the Meiji Restoration.

The defeat in World War II, therefore, did not create the conflicts I have been describing but rather intensified them. Yet the intensification has been of a very special kind, adding important new dimensions to the postwar situation. Most important here was the humiliation of the defeat itself, because in that defeat Japan experienced not only its first great modern "failure" (after a series of extraordinary successes) but also had its mystical-ideological concept of *kokutai* undermined. *Kokutai*[6] is usually translated as "national polity" or "national essence," but it also conveys the sense of "body" or "substance," and its nature is impossible to define precisely. Included in *kokutai* are the concepts of "national structure," particularly the emperor system; "national basis," the myth of the divine origin of Japan and of its imperial dynasty; and "national character," those special Japanese moral virtues, stemming from both native and Confucian influences, that are considered indispensable for individual behavior and social cohesion (embodied in *Bushidō,* or the Way of the Warrior). Although *kokutai* is a relatively modern concept—manipulated for political purposes during the Meiji era and again in association with pre-World War II militarism—it had profound roots in Japanese cultural experience and embraced something in the cultural identity of all Japanese.

Most young people (with the exception of "rightists") no longer take *kokutai* seriously; they dismiss it as the propaganda of militarists, and even find it laughable. Nevertheless, the dishonoring of *kokutai* has created in many Japanese youth a sense of their own past as dishonored, or even of Japaneseness itself as dishonored. The sudden collapse of *kokutai* revealed its tenuousness as an ideological system. But it also created an ideological void and thus encouraged the polarizing tendencies that still haunt Japanese thought —the urge to recover *kokutai* and make things just as they were, and the opposite urge to break away entirely from every remnant of *kokutai* and make all things new.

Nor can intellectual youth feel comforted by Japan's extraordinary postwar industrial development. As the first generation of Asians to grow up in a country which, at least in its urban aspects, resembles the modern industrial West, they are also the first to experience the dehumanizing effects of mass society (though they sometimes attribute these to capitalism alone). Moreover, they link this industrial development to the "old guard" among their politicians and businessmen, from whom they feel themselves (or wish to feel themselves) completely removed. They find insufficient satisfaction in the democratic freedoms they enjoy—they often do not *feel* free—and they condemn themselves for being attracted to the rewards of their own society.

The collapse of *kokutai* also ushered in a new era of increased receptivity to outside ideological currents. But when young intellectuals now look to the West, they find the Western world itself in a state of profound uncertainty and disillusionment in relation to much of its own great tradition of humanism, individualism, Judeo-Christian religion, and private economic enterprise. They see in Communism a powerful, expanding force, with profound intellectual, emotional, and moral attractions (especially in the case of Chinese Communism), but they have been sufficiently sensitive to the organizational cruelties of Communism for much disillusionment to have set in here as well. Still inspiring and untarnished in their eyes is the social revolution occurring throughout most of Africa and Asia (and in other relatively underdeveloped areas, such as Latin America), whose dynamism has great appeal. But the youth are inwardly torn between their "Asian" identification with this movement and their "Western" separation from it—that is, by the experience of Western-inspired "modernization," which (superimposed on their previous geographical and cultural isolation) has set the Japanese apart from the rest of Asia and has enabled them to accomplish many of the things other Asian countries are just now setting out to achieve.

Surely, it is not only in postwar Japan that such a break in the sense of connection has occurred. To what extent can we say that universal factors are at play? Here we must first consider the ever-present ideological gap of the generations, found in varying degrees in all cultures and at all periods of history. Thus, Ortega y Gasset claims that "the concept of the generation is the most important one in the whole of history."[7] He points out that the twenty-year-old, the forty-year-old, and the sixty-year-old create three different styles of life which are blended into one historical period, so that "lodged together in a single external and chronological fragment of time are three different and vital times." Ortega y Gasset calls this "history's essential anachronism," an "internal lack of equilibrium," thanks to which "history moves, changes, wheels and flows." In other words, this generational gap is the psychobiological substrate of the historical process, imperfectly blended with it but necessary to it. Moreover, the occurrence of "youth problems" and "youth rebellions" throughout the world suggests that the gap is universally enlarging. The rapid technological and social change affecting all mankind has created a universally shared sense that the past experience of older generations is an increasingly unreliable guide for young people in their efforts to imagine the future. And individual identity diffusion becomes for many young people everywhere a virtual necessity, a form of sensitive (though often costly) experimentation with historical possibilities. In Japanese youth, cultural and historical influences have brought about diffusion and dislocation of unusual magnitude.

Selfhood

One of the ways in which young people attempt to deal with this predicament is by stressing a developing awareness of their own being, by delineating the self. They do this in many different ways. They speak much of individual freedom in relation to family and society, and strongly criticize the negation of the individual person in traditional (and contemporary) Japanese practice. They respond strongly to those elements of Marxist thought which refer to self-realization. And they frequently combine their Marxism with existentialism, for they are drawn to the ideal of personal freedom they find expressed in the writings of Jean-Paul Sartre and in his life as well. They criticize great nations like Russia and America for what they perceive to be a tendency toward mass conformity and a denial of self. And many criticize their own student political movement on the same basis, despite strong sympathy for it otherwise. Still others conduct their self-exploration through an attitude of negation, through the mood of nihilism and passive disintegration that has frequently appeared in Japanese literature and social behavior; students have this kind of attitude in mind in their use of a coined word meaning "feigned evil" *(giaku)*.

Also related to this urge to liberate the self is the extremely widespread fascination, even among intellectual youth, with American Western films. Both Sato and Kondo attend them regularly, and have revealed a variety of reasons for their appeal: the exhilarating spectacle of young men and women engaged in purposeful adventure, free from conventional pressures of social obligation *(giri)*, and creating a new way of life solely by their own efforts; the sense of geographical openness and of unlimited possibility; the admirable figure of the hero—his simple courage, direct (unambivalent) action, and tight-lipped masculinity; and the excitement and precision of the gunplay. All this, of course, is contrasted with their own situation in present-day Japan. They perceive in "Westerns" a world of ultimate freedom, in which the self is clearly defined, unrestrained, and noble even in its violence.

But underneath this ideal of selfhood, however strongly maintained, one can frequently detect an even more profound craving for renewed group life, for solidarity, even for the chance to "melt" completely into a small group, a professional organization, or a mass movement, and even at the cost of nearly all self-assertion. Those most concerned with selfhood have often told me (as Kondo did) that their moments of greatest happiness come when they feel themselves, in a spiritual sense, completely merged with groups of young comrades. And I have repeatedly observed their despair and

227

depression when separated from groups with which they have been profoundly involved, or when unable to establish meaningful group relationships. For Japanese of all ages, in virtually any situation, have a powerful urge toward group formation: when they wish to do something startling (intellectual, artistic, social, or political), they are likely to go about it by forming, joining, or activating a group. The extraordinary array of student circles, of cultural, professional, political, and neighborhood groups—the "horizontal" groups so prominent at all levels of society—makes Japan one of the most group-conscious nations in the world.

One feels this tension between the ideal of individualism and the need for the group in the concern of young people with that much-discussed, elusive, sometimes near-mystical, but always highly desirable entity known as *shutaisei*. *Shutaisei* literally means "subjecthood," and is a modern Japanese word derived from German philosophy, coined by Japanese philosophers to introduce into Japanese thought the German philosophical ideal of man as subject rather than object. But the word has had its vicissitudes: some philosophers who were sympathetic to the prewar Japanese ideology sought to combine *shutaisei* with *kokutai* (thereby almost reversing its original meaning); while in the postwar period it has been a central concept in intra-Marxist debates about man's nature and responsibility in relation to the historical process. Continuing in this postwar trend, young people use *shutaisei* to mean two things: first, holding and living by personal convictions—here *shutaisei* comes close to meaning selfhood; and second, having the capacity to act in a way that is effective in furthering historical goals, and (at least by implication) joining forces with like-minded people in order to do so—here the word means something like social commitment. The young Japanese themselves tend to be confused by the conflicts which seem to arise from these two aspects of *shutaisei*. Their greatest difficulty is in realizing to their own satisfaction its first element, that of selfhood; and the sense of "smallness" or the "inferiority complex" which they talk so much about seems to reflect the great difficulty the Japanese have in perceiving and believing in a relatively independent self.

Yet the very groups to which youth are drawn may themselves become arenas for the struggle for selfhood. In this group life there is always a delicate balance between competition (often fierce, though usually suppressed), mutual support and encouragement, and (perhaps most important) constant comparison with other members as a means of self-definition. Moreover, their struggle for selfhood in combination with their historical dislocation has resulted in a burst of literary and artistic creativity among them; such creative accomplishments rarely resolve their dilemmas, but they are energetic efforts to come to grips with them.

Traditional Japanese patterns of group and individual behavior throw a good deal of light upon the present situation. For, in Japan, the stress upon the group as the "cellular unit"[8] of society and the negation of consciousness of self has been carried to an unparalleled extreme. The relatively closed ("vertical") groups constituting traditional society (family, locality, clan, and nation) became the source of all authority in Japanese life. (An outspoken critic of *Tokugawa* society characterized its group hierarchy as "tens of millions of people enclosed in tens of millions of boxes separated by millions of walls."[9]) More than this, these groups have often become something close to objects of worship, resulting in a characteristic Japanese pattern which we may term *deification of the human matrix*.[10] In a culture with a notable absence of universal principles, men have found their sacred cause in defending the integrity of their particular human matrix.

This pattern finds its central symbolization in the idea of the divinity of the Imperial family, the Imperial family having been considered since ancient times the "living totality of the nation." It found later expression in the *samurai's* absolute submission and loyalty to his feudal overlord, which in turn supplied a psychological model for the modern practice of national Emperor worship. Therefore, it was not simply the experience of *Tokugawa* isolation which brought about the cliquish intensity of Japanese group life, as is often asserted; rather it was the earlier tendency to regard the Japanese human matrix as sacred which created (and then was reinforced by) *Tokugawa* isolation. The Japanese have long had an unconscious tendency to equate separation from the human matrix (or exile) with death—expressed linguistically in the common Chinese character used in the Japanese words *bōmei suru* (to be exiled, literally, to lose life), and *nakunaru* (to die).

All this has resulted in a language and a thought pattern in which "there is . . . no full awareness of the individual, or of an independent performer of actions . . . no inclination to attribute actions to a specific performer."[11] It is the combination of historical inability to delineate boundaries of the self with the modern urge to do so that creates the inner conflicts I have described.

This historical tendency has its counterparts—indeed, they are partly its results—in child-rearing practices and in resulting individual psychological patterns. Japanese children in relation to their parents (and especially to their mothers) are expected to show the desire to *amaeru*. *Amaeru* has no single English equivalent. It means to depend upon, expect, presume upon, even solicit, another's love. According to Dr. Takeo Doi, this pattern of *amaeru*—or *amae*, the noun form (both words are derived from *amai*, meaning sweet)— is basic to individual Japanese psychology and is carried over into adult life and into all human relationships. Doi argues further that the unsatisfied urge to *amaeru* is the underlying dynamic of neurosis in Japan.[12] We can also say that the *amaeru* pattern is the child's introduction to Japanese group life; brought up to depend totally on his mother (and to a lesser extent his father and older brothers and sisters within the family group), he unconsciously seeks similar opportunities later on in relation to others who are important to his welfare. The lesson he learns is: you must depend on others, and they must take care of you. It is difficult for him to feel independent, or even to separate his own sense of self from those who care for him or have cared for him in the past. The spirit of *amaeru* still dominates child-rearing practices; and the desire expressed by many young Japanese (especially women) to bring up their own children "differently," more "as individuals," is a form of recognizing that this spirit conflicts with aspirations to selfhood.

The question arises whether there is in the Japanese historical and cultural tradition a tendency opposite to those we have mentioned, one stressing greater independence, more self-expression, and less submission to group authority. Many Japanese feel that such a native tendency did exist before it was submerged by the repressive atmosphere of Confucian orthodoxy, and they point to such early Japanese writings as the *Manyoshu* (a collection of verse recorded during the seventh and eighth centuries) as depicting considerable spontaneity of emotion and independence of spirit. This is a question I will not attempt to take up; but I believe one can say that, despite this early ethos of spontaneity, there remains a rather weak tradition for the ideal of individuation which young people now embrace.

For the concepts of selfhood and of commitment (which implies selfhood), of *shutaisei*, stem almost entirely from Western tradition: from classical Greece of the fourth century B.C., from Judeo-

Christian monotheism, from the Renaissance and the Reformation, and from the later philosophical schools which grew out of these traditions. Vital to these Western traditions has been a spirit of universalism—concepts of the universal God, the universal Idea, and the universal State—with which the individual self could come into nourishing symbolic contact and thereby free itself from the influence of more immediate and particularistic human groups. This kind of stress upon individuation has been limited primarily to a very small part of the world, mostly Western Europe and those areas populated by Western Europeans. Non-Western cultures have been profoundly stirred by it, but have invariably found it necessary to make a compromise with it, to adopt a form of *self-expression via the group*. The alternative is a retreat from selfhood into a modern form of collectivism that makes emotional contact with earlier group traditions.

It is perhaps unnecessary to add that the process of individuation has hardly run smoothly and is far from "complete" in those Western cultures where it evolved. Moreover, the exciting appeal of selfhood has created, first in the West and then in those countries influenced by the West, what we might term a myth of absolute individualism, the fantasy of the self existing in total independence from all groups. This, of course, ignores the psychological interdependence between individuation and community.

The dual aspects of *shutaisei* reflect an awareness in young Japanese of the need for both. But they cannot achieve either aspect of *shutaisei* without modifying their overwhelming need for immediate group acceptance, since this can stifle not only selfhood but true social commitment as well. Their inner question is not so much, "Who am I?" (the problem of identity) as, "Can I perceive my own person as existing with a measure of independence from others?" (the problem of selfhood).[13]

Logic and Beauty

Young Japanese repeatedly assert their desire to be logical, objective, scientific, to be in every way tough-minded. They stress their urge to *warikiru*—a verb which means "to divide" but which now conveys the sense of cutting through a problem, giving a clear-cut and logical explanation. They are quick to criticize one another's attitudes as *amai* ("sweet"), meaning wishful, rather than realistic (*genjitsuteki*) in one's expectations; or as *kannenteki*, meaning prone to philosophical idealism, overtheoretical, and also unrealistic. A still stronger condemnation is *nansensu* (nonsense) which has been used particularly widely within the student movement to dismiss dissenting opinions and convey the sense that such opinions are a logical impossibility.

These "undesirable" (and "illogical") tendencies are in turn related to one's background and probable future: those judged guilty of them are likely also to be called *botchan*, meaning "little man" or "sonny," and suggesting the softness and self-indulgence created by favored middle-class circumstances; *puchi-buru*, the Japanization of petty bourgeois, which conveys utter contempt in its very sound; or *sarariman* (salaried man), the *bête noire* of Japanese youth in the wider social sense, signifying the selfless, mindless, amoral, modern Japanese automaton, whose thought and life are utterly devoid of "logic" (and which, it must be added, most young Japanese expect to become).

Yet accompanying this strongly held ideal of logic is an inherent predilection for nonrational, aesthetic responses, and this predilection becomes increasingly evident the better one gets to know a young person. Dedicated political activists have told me that they were inspired to join the revolutionary movement through the examples of the heroes of novels by Gorky, Rolland, and Malraux. One *Zengakuren* leader, a central figure in the mass political demonstrations of 1960, told me he had been "profoundly moved" (*sugoku kangeki shita*) by the "absolute sincerity" (*shinjō o tsukusu*) of the revolutionaries in Gorky's novel *The Mother*, by their capacity to hate their exploiters and at the same time to love one another, and that he later found his fellow *Zengakuren* activists to be in the same way inwardly "beautiful" (*utsukushii*). Such sentiments were also prominent in Sato and Kondo.

This emphasis upon the sincere (*seijitsu*) and the pure (*junsui*) applies to not only politics but to all experience. And among the majority of students (those who are politically moderate or apolitical) there is often a guilty sense of their being unable to match the "purity" and "sincerity" of *Zengakuren* leaders, despite their feeling otherwise critical of their behavior. Many speak of their desire to live seriously and honestly; the word they use is *shinken*, whose literal meaning is "true sword," and which suggests the kind of inner intensity one might find in art or religion or in any dedicated life.

When talking freely, they make extensive use of the rich Japa-

232

nese vocabulary of aesthetic and emotional experience—*kimochi* (feel), *kibun* (mood), *kanji* (feeling), *kankaku* (sense), *kanjo* (emotion or passion), *kan* (intuitive sense), *funiki* (atmosphere) etc.; and *akarui* (bright) or *kurai* (dark), to convey their impression of almost any event or person. They have an unusually strong aesthetic response to the totality of a situation, and both their immediate and enduring judgments depend greatly on the extent to which purity and beauty are perceived. I believe that one of the reasons for the attraction of Marxism as an over-all doctrinal system is its capacity to evoke this sense of aesthetic totality, of the universal "fit" and feeling of truth. At the same time, Marxism readily lends itself to the equally necessary stress upon "scientific logic" and tough-minded analysis.

The reliance upon aesthetic emotions extends into personal relations, too, in which one finds that these students combine a considerable amount of distrust and criticism (especially toward their elders) with a profound romanticism, a strong tendency to idealize human emotions. Their "wet" (in the slang of postwar youth) quality is especially evident when falling in love: the young man, perhaps after speaking a few words to a young woman in a casual, more or less public situation, or perhaps after simply catching a glimpse of her, sends an impassioned letter declaring his love and accepting the responsibility (meaning his readiness to consider marriage) for having done so; and in the relationship which follows, the letters exchanged (almost entirely devoted to descriptions of feeling and mood) often seem to be of greater importance than the rare meetings for talks in coffee houses. To be sure, there are more "modern" ("dry") relationships also, but it is surprising how frequently one still encounters this older "Japanese" form of love affair among students.[14]

The efforts to resolve this tension between aesthetic emotion and logic sometimes result in rather problematical attitudes toward ideas in general. At one extreme is the desperate urge to cast off the alien "logic"—the distrust of all ideas, theory, or even talk, and the stress on the pure and spontaneous (aesthetically perfect) act. This pattern is most intense in political rightists, who are rare among intellectual youth, and in certain postwar literary movements whose leaders have emerged as university students, expressing disdain for the intellect in favor of a cult of the senses.

The opposite tendency, and the more frequent one, is to elevate logical and scientific ideas to the status of absolute, concrete entities, which then take on aesthetically satisfying properties and become incontestable—a form of scientism. But one must add that many show great sensitivity in groping toward a balance between their logical and aesthetic inclinations, allowing themselves an increasing capacity for precision and logic while neither disdaining the nonrational nor worshipping a pseudo-scientific form of rationalism.

These conflicts become more understandable when we consider that Japanese youth are heirs to a tradition utterly unique among high cultures in its extraordinary emphasis upon aesthetic experience and the neglect of logical principles. The aesthetic emphasis includes not only a remarkable body of art and literature but also a consistent concern with sensitivity to all varieties of beauty and every nuance of human emotion. Such aesthetic sensibility can even become the criterion for human goodness, as suggested by Motoori Norinaga, a leading figure in the eighteenth-century Shinto revival, when commenting (approvingly) upon the morality of *The Tale of Genji* (the great court novel of the early eleventh century): [15]

Generally speaking, those who know the meaning of the sorrow of human existence, i.e., those who are in sympathy and in harmony with human sentiments, are regarded as good: and those who are not aware of the poignancy of human existence, i.e., those who are not in sympathy and not in harmony with human sentiments, are regarded as bad.

There is a corresponding stress, found in almost every Japanese form of spiritual-physical discipline (Zen *jūdō, kendō, karate*), on achieving emotional harmony, purity, and simplicity: a form of aesthetic perfection in which conflict (ambivalence) is eliminated. But among the "impurities" and "complexities" got rid of are ideas and rational principles. Again, Motoori Norinaga: [16]

In ancient times in our land, even the "Way" was not talked about at all and we had only ways directly leading to things themselves, while in foreign countries it is the custom to entertain and to talk about many different doctrines, about principles of things, this "Way" or that "Way." The Emperor's land in ancient times had not such theories or doctrines whatever, but we enjoyed peace and order then, and the descendants of the Sun Goddess have consecutively succeeded to the throne.

Averse to detailed general principles, the Japanese have tended to turn to their opposite, to brief, concrete, emotionally evocative symbols, as in the short verse forms of *Tanka* and *Haiku*. In the political-ideological sphere, however, this propensity for evocative

234

word-symbols, for what one Japanese philosopher has called the "amuletic" use of words,[17] has had more serious consequences. Words like *kokutai, nipponteki* (Japanese) and *Kōdō* (Imperial Way) seem to have given their users a magical sense of perfection; and this same tendency has made Japanese particularly susceptible to slogans associated with military expansionism. These amuletic words and slogans, within the framework of *kokutai*, offer a sense of aesthetic totality, of both moral righteousness and group invulnerability. And the Japanese language itself reflects the tendencies we have mentioned in its unusual capacity for describing beauty and capturing emotional nuance and in its contrasting limitations in dealing with precise ideas—such that it "has a structure unfit for expressing logical conceptions."[18]

When Japanese students condemn the "irrationalism" of their tradition, they are repeating the attitudes of generations before them, going back to the middle of the nineteenth century; and even today intellectuals of all ages feel themselves to be (rightly or wrongly) an island of logic amid a sea of emotionalism. Their stress on logic represents a cultural countertrend, a rebound reaction against a tradition which has not only neglected scientific and rational thought but also has often condemned it. Such a cultural sequence can readily lead to a form of worship for the thing that has been historically denied, and this makes it difficult for many young Japanese to retain full access to the aesthetic sensitivities still at the core of their self-process.

Turning to the universal aspect of the question, we find that the Japanese have been unique only in the *degree* of their stress upon aesthetics and neglect of logic. Their style of symbol-formation (following the terminology of Susanne Langer)[19] has stressed "nondiscursive" elements which rely upon a "total," or essentially emotional and aesthetic, form of reference, in contrast to the relatively great stress in the West on "discursive" (or logical) symbolic forms. But advanced logical skills are a relatively late accomplishment, in a historical as well as an individual psychological sense, and are always superimposed upon an earlier, nonrational, mental structure. Moreover, the high development of logical thought has created in modern Western man an artificial separation of mind into logical and nonlogical categories. The glorification of the former and the derogation of the latter has left men dissatisfied with the myths and symbols they used to find enriching, although it hardly seems to have eliminated their irrationality. It can be said that this separation of mind has been the price Western man has had to pay for his modern achievements, including that of selfhood.

Japanese intellectuals, especially young ones, are now seeking similar achievements and paying a similar price. Yet Western man has had some second thoughts about the matter, not only in spiritual quests but also in the realization that original discoveries in such logical disciplines as mathematics and the physical sciences depend importantly on aesthetic and other nonrational experience. In the same light, one suspects that the recent emergence of gifted Japanese mathematicians and scientists reflects not only the rapid development of logical thought in post-Meiji Japan but also the capacity of outstanding Japanese to bring to bear upon their intellectual work elements of their exceptionally rich aesthetic tradition.

Directions and Principles

It is clear by now that the psychological directions in which young people in Japan are moving (and in which Japan itself is perhaps moving) are spasmodic, conflicting, and paradoxical. Yet we can discern reasonably definite patterns and crucial pitfalls.

There is first the conscious ideal, the symbolic direction which a large portion of young people chart out for their own character structure. They wish to be, and are to some extent becoming: "new," progressive, innovative, and antitraditional; active, individually independent and socially committed (possessors of *shutaisei*); logical, realistic, tough-minded, and scientific. Summing up the spirit of this ideal path, we may call it "active-Western-masculine."

In opposition to this direction is their negative image,[20] the things they wish to avoid becoming, but which deep in their mental life (because of their individual and cultural experience) in many important ways, they *are*. They do not want to be: "old," unprogressive, acquiescent, traditionalistic; passive, wholly dependent upon others and socially uncommitted (lacking *shutaisei*); irrational, unrealistic, wishful and unscientific. This path we may term (according to the way they perceive it) "passive-Japanese-feminine."

The inner struggle between these two sets of elements is continuous. The actual psychological task—and one which Japanese have been performing since the seventh century[21]—is that of making use of "old" tendencies when creating new patterns, always careful that the new patterns do not do violence to what is emotionally most important in the old tendencies. Thus young Japanese are seeking to recover their own past even as they move away from it; to maintain their sense of group intimacy, even as they achieve greater individuation; to live by their aesthetic sensitivities, even as they attain greater logical precision. In outstanding youth, much of this process is conscious and self-evaluative, while in the majority it is largely unconscious. In all, the distinction between what is "ideal" and what is "negative" becomes less absolute as it is found necessary to make use of both sets of elements. And even the most mentally adventurous can at most achieve only a slight modification of the psychological patterns which form the core of their developing self-process. But that is achieving a great deal.

The two great problems young people face are, on the one hand, totalism (or psychological extremism),[22] and on the other, a complete surrender of ideals, or what might be called "moral backsliding." They themselves articulate the second problem clearly, the first more vaguely.

Totalism may take (and has taken) two forms. There is the *totalism of the new,* in which young people carry the elements of the "active-Western-masculine" ideal to the point of creating a closed ideological system (usually derived from Marxism) in which there are combined elements of extreme idealism, scientism, a moral imperative for bold (sometimes violent) action, and a degree of martyrdom. Such totalism has been prominent in the student movement, although actively espoused by only a small minority; and even among this minority, many later struggle against their totalism by questioning the "openness" of the ideology and the nature of their own involvement in it.

The *totalism of the old* disdains the symbols of logic, science, and *shutaisei,* in favor of a traditionalistic reversion to *kokutai;* violent homicidal assaults are made by young rightist fanatics in order to "protect" and "restore" the sacred Japanese identity. Among university students, one encounters, rather than the overt expression of this form of totalism, a certain amount of covert identification with it on the basis of the rightists' youthfulness, sincerity of feeling, and purity of motive.

We have already mentioned some of the elements in Japanese culture which lend themselves to totalism—patterns of absolute self-negation and exaggerated dependency in relation to a deified human matrix. But there are also important cultural elements which resist totalism: a long-standing tolerance toward diverse influences from the outside; a distrust for things which are immoderate or forced (*muri*); a general acceptance of bodily functions and an attraction to the sensual pleasures of this "floating world" (*ukiyo*); and the more recent disillusionment with chauvinistic nationalism and an attraction toward (if only partial comprehension of) moderate democratic patterns. Moreover, what G. B. Sansom refers to as the great sensitivity of the Japanese to the surfaces of existence tends to give a hysterical rather than a profound or totalistic quality to Japanese expressions of extremism.

The second great problem, that of "moral backsliding," involves giving up one's ideals in order to make one's peace with organized society (as we saw in the case of Kondo), and reflects a long-standing Japanese tendency known as *tenkō*. *Tenkō* means conversion, a form of about-face, and usually suggests surrendering one's integrity in order to merge with a greater power. In this sense, every youth is expected to have, and every youth expects to have, an experience of *tenkō* on graduating from his university and "entering society," and this is sometimes compared with the *tenkō* of intellectuals who gave up Marxist and democratic ideals during the 1930's to embrace some version of *kokutai*. Thus *tenkō* is basic to Japanese psychology: it reflects patterns of aesthetic romanticism, obscurantism, and often shallow experimentation with ideals prior to *tenkō* itself; and it also reflects the ultimate need felt by most Japanese to submit and become part of existing authority, to gain a safe place in a human matrix, rather than risk standing alone. Young Japanese go back on their ideals because their society virtually forces them to; but their own emotional inclinations contribute to this "self-betrayal."

Yet much of what I have described may be understood as youth's efforts to resist *tenkō* and to acquire a new form of integrity. Without and within, their struggle is no easy one. But they bring to it a special intensity that has long characterized their culture.

It is their intensity which helps one, when working with Japanese youth, to think more vividly about general principles. And from this study I believe it is possible to attempt a few general formulations on the relationship between youth and history.

Historical change is not in its most elemental sense brought about by "technology" or "science"; it is human in origin, a product of creative and destructive expressions of the human organism. We have said that the gap between the generations supplies a basic psychological substrate for historical change. But there would be no generational gap—that is, every son would think, feel and do as his father did—were it not for another psychological tendency: the inherent urge toward exploration and change, which is part of the growth process and exists side by side with the more conservative human tendency to hold on to old emotional patterns and adapt to things as they are. The urge toward change becomes closely linked with man's efforts to master his physical environment; it is the source of the ideas, discoveries, and technologies, which in themselves exert so great an effect in reshaping man's experiential world that their own human origins are obscured and often forgotten. The more rapid the over-all process, the greater the generational gap—or, one might say, the shorter the time required for a "new generation" (a youth group with new attitudes) to appear.

Young people in their late teens and early twenties are central to this process, not because they make great historical decisions or discover great truths, but because they feel most intensely the generational gap and the inner urge toward change. At an age when self must be created or defined and identity discovered, their strong response to ideals and ideologies produces pressures toward change —sometimes constant, sometimes explosive—in all societies. (I suspect that even in "primitive" or in "static" societies, careful historical study would reveal evidence of the generational gap and the inherent urge toward change, however these may have been suppressed by techniques of "initiation" into existing patterns of adult life.) In our present era social change has been so rapid and the effects of the "second scientific revolution" so momentous that we are in the midst of universal historical dislocation. Consequently, youth experiences traditional symbols painfully; and the family representative of traditional authority, the father, tends to "disappear." Youth bands together, partly in an effort to set its own standards, partly to experience collectively ideologies that promise to bring about further change felt to be imperative, and partly in the (less conscious) effort to recover something from its own past that might lead to greater stability.

Historical change in all cultures, except those which are the vanguard of such change (now those of America and Russia), depends very greatly on outside influences, particularly on influences coming from the vanguard cultures. Gifted young people are extremely sensitive to these outside influences but are also ambivalent to them. They are attracted to their liberating elements and at the same time are fearful of having their own cultural identity overwhelmed; and their ambivalence can lead to sudden shifts from near-total embrace to near-phobic avoidance. Moreover, outward acceptance can be a means of maintaining a deeper resistance. Yet even when this is the case, these outside influences do bring about gradual changes in individual psychology. Such changes become significant when core elements of self-process within a particular culture are permitted expression in the new combination taking place.

Finally, youth's intensity in relation to the historical process gives it a particularly strong potential for totalism. This potential is most likely to be realized when young people feel hopelessly dislocated in the face of rapid and undigested historical change, and when they are convinced that their society will afford them no recognition without moral backsliding. But the capacity of young people for self-examination within a social and historical framework—sometimes exasperating in its seeming narcissism—can be an effective source of resistance to such totalism. It can also open up new possibilities in the desperate universal task of coming to grips with the ever accelerating, ever more threatening movement of history.

REFERENCES

1. This study of Japanese youth, supported by the Foundations' Fund for Research in Psychiatry, is still in progress. The research subjects I have interviewed are largely an elite group, attending leading universities and women's colleges in Tokyo and Kyoto, and in many cases possessing outstanding abilities as students and student leaders. I am grateful to Dr. Takeo Doi, with whom I have consulted regularly during the work; and to Mr. Hiroshi Makino and Miss Kyoko Komatsu for their general research assistance, including interpreting and translation.

2. This in many ways resembles the "apparently closed room" which Paul Goodman describes as confronting American youth (*Growing Up Absurd,* New York, Random House, 1960). But rather than the "rat race" which American youth encounter, Japanese youth are more concerned with their society's stress upon one's place or slot, which, once assigned, is (at least, occupationally) difficult to change.

3. Nowadays, more than half the students admitted to Tokyo and Kyoto Universities have spent at least one year as a *rōnin.* Because of the prestige and better job opportunities accorded graduates of these leading universities, students prefer to spend an extra year (or sometimes two or three years) working to gain admission, rather than attend a different university.

4. Ryusaku Tsunoda, William Theodore de Bary, and Donald Keene, *Sources of Japanese Tradition* (New York, Columbia University Press, 1958), pp. 592 and 606.

5. I do not mean to suggest that this modern historical experience is the only cause of the Japanese tendency to compartmentalize their beliefs and identifications. Moreover, there is a good deal of evidence that the same tendency existed during the seventh, eighth and ninth centuries in relation to Chinese cultural influences. See Ruth Benedict's discussions, "Clearing One's Name" and "The Dilemma of Virtue," in *The Chrysanthemum and the Sword*, Boston, Houghton Mifflin Company, 1946.

6. For discussions of *kokutai*, see Maruyama Masao, "Chōkokka-shugi no Ronri to Shinri" (Theory and Psychology of Ultranationalism), in *Gendai Seiji no Shisō to Kōdō* (Thought and Action in Current Politics), Tokyo, 1956; Tsunoda, de Bary and Keene, *op. cit.*, 597-598; Richard Storry, *The Double Patriots* (Boston, Houghton Mifflin Company, 1957), p. 5; and Ivan Morris, *Nationalism and the Right Wing in Japan*, London, Oxford University Press, 1960. I am also indebted to Professor Maruyama for personal discussions of *kokutai* and *shutaisei*.

7. José Ortega y Gasset, *What is Philosophy?* (New York, W. W. Norton & Company, 1960), pp. 32-39.

8. Kawashima Takeyoshi, "Giri," *Shisō* (Thought), September, 1951, as quoted in Nobutaka Ike, *Japanese Politics* (New York, Alfred A. Knopf, 1957), p. 29.

9. Fukuzawa Yukichi, as quoted by Maruyama Masao, "Kaikoku" (The Opening of the Country) in *Kōza Gendai Rinri* (Modern Ethics), Tokyo, Chikuma Shobō, 1959.

10. Much of the following discussion is based upon Nakamura Hajime, *The Ways of Thinking of Eastern Peoples* (Tokyo, Japanese National Commission for UNESCO, 1960), pp. 304-433.

11. *Ibid.*, p. 307.

12. Doi Takeo, "Jibun to Amaeru no Seishin Byōri" (The Psychopathology of the Self and Amaeru), *Seishin Shinkei Gaku Zasshi* (Journal of Neuropsychiatry), 1960, *61:* 149-162; and "*Amae*—A Key Concept for Understanding Japanese Personality Structure," (unpublished manuscript). Dr. Doi emphasizes correctly, I believe, that the emotions surrounding *amaeru* are by no means unique to the Japanese but are particularly intense in them.

13. This idea of "self" is closely related to that of Susanne Langer, who states, "The conception of 'self' . . . may possibly depend on this process of symbolically epitomizing our feelings" *Philosophy in a New Key* (New York, Mentor Books, 1948), p. 111; and also that of Robert E. Nixon: "Self is the person's symbol for his own organism" in "An Approach to the Dynamics of Growth in Adolescence," *Psychiatry*, 1961, *24:* 18-31.

14. In looking toward marriage, most students express a strong preference for "love marriages" rather than the more traditional "arranged marriages." But despite this general trend, a considerable number, when the time comes, resort to the older pattern of family arrangements.

15. Tsunoda, de Bary, and Keene, *op. cit.*, p. 533.

16. Nakamura, *op. cit.*, p. 471.

17. Tsurumi Shunsuke, as quoted and summarized in Morris, *op. cit.*, Appendix I, pp. 427-428.

18. Nakamura, *op. cit.*, p. 465.

19. Langer, *op. cit.*, pp. 75-94. I would stress that this is a *relative* difference in emphasis between Japanese and Western patterns of symbolization.

20. This "negative image" is closely related to Erik H. Erikson's concept of "negative identity," except that it is not, as in the case of the latter, something that youth have been warned not to become, but rather a part of their culture within themselves which they condemn. The general point of view in this article is influenced by Erikson's psychological approach to historical events, especially *Young Man Luther*, New York, W. W. Norton & Co., 1958. See also Zevedei Barbu, *Problems of Historical Psychology*, London, Routledge & Kegan Paul, 1960.

21. Nakamura, *op. cit.;* and Yoshikawa Kōjiro, "The Introduction of Chinese Culture," *Japan Quarterly*, 1961, 8: 160-169.

22. I have discussed this tendency at greater length in my book, *Thought Reform and the Psychology of Totalism*, New York, W. W. Norton & Company, 1961.

LAURENCE WYLIE

Youth in France and the United States

LIFE IS DIFFERENT for French and American adolescents. The differences, of course, are less obvious than those between widely separated, primitive cultures, and they are complicated by the social and geographical variations within modern, national cultures. In certain respects a New York boy in a college preparatory school may have more in common with a Parisian in a *lycée* than with a Negro boy in Harlem. The son of dairy farmer in Normandy may have a great deal in common with a boy on a farm in Wisconsin and very little with a boy in the coal country of northern France. Nevertheless, there is something about the adolescent experience of a French boy which makes it different from the normal adolescent experience of an American. The basic differences which exist between French and American adolescents transcend social class and are evidence of contrasts between the cultures themselves.

One way in which Americans and Frenchmen seem clearly to differ is in their conception of the rules that govern social behavior. The French generally believe that it is right for people to be forced to accept the sharply defined framework which man has projected onto the chaos into which he is born. Americans, on the other hand, generally feel that individuals should not be hampered in their free development but should discover for themselves the rules that govern the naturally ordered reality into which they are born. Taking these two attitudes as starting points, we can trace their effects on French and American adolescents as they face their common problems—the problem of learning to live with physically maturing bodies, the problem of trying to fulfil the image they have formed of themselves as adults, the problem of preparing to share in the responsibility of perpetuating their culture by having and educating children.

243

French children are taught very early the importance of man-made limits. A few years ago in an interview with a young French woman living in this country, I doubted that we could accomplish our business during the short time at our disposal when I realized that Madame's two-year-old son was obviously to be on hand for our discussion. My fears were unfounded, however. The child was put on the floor in a corner of the room with one toy. In front of him his mother put a cardboard box and said: "Pierre, there's a line running through the box. Your place is on that side of the line. Ours is on this side." To my amazement, the child respected the line which had been projected across the corner of the rug, and not until his mother and I had finished our business and she took away the box did he venture out from what had been for a few moments his own well-defined segment of existence.

From the time he is very small the French child learns that life is compartmentalized. The limits defining the compartments, furthermore, are established by forces beyond his control. Pierre was not consulted about whether he wanted to stay in one part of the room or another. He was not asked where he wanted the line drawn or indeed whether he wanted a line at all. Even the fact that the line was invisible was of no importance: this imaginary barrier was as forbidding as a wall of wood and plaster.

A French child grows up with an awareness of both kinds of barriers, one a material barrier defining our relationship with the physical world, the other a social barrier defining our relationship with people. He learns, furthermore, that each segment of existence calls for behavior appropriate to it. When he goes to school he continues to learn in the same compartmentalizing manner. He learns by rote, for example, the categories of history and geography and grammar that have been established by someone else—the authors of the textbooks or his teachers—and he then studies examples of these categories until he can recognize them by himself. Learning is essentially a matter of acquiring a clear awareness of the compartments of existence, of their distinctiveness, of their interrelationships.

In contrast to Pierre, most American children I know are being brought up in a quite different manner. The most extreme case I think of is that of a psychologist and his wife who as a matter of principle raised their child to be unaware of previously established limits. The child had no playpen. He could crawl about the house at will, which meant that an adult had to be with him every minute to make sure he did not hurt himself. He was never told that he must not do anything. He was offered alternatives, but he was not to hear the word "no." There was no special time or place for feeding, sleeping or toilet. In so far as it is possible, this child was supposed to establish his own limits as he explored the world.

This may be an extreme case of American permissiveness, but in tendency it is by no means uncommon. In fact, even the sort of training I had as a child fifty years ago, although far from permissive, was basically more like that of the psychologist's child than Pierre's. It seemed to me as I was growing up that I was surrounded by many rules and limits—"get to school on time"—"no skating or noisy play on Sunday"—"wash hands before meals"—"don't shock the women of the Ladies Aid Society." These were human rules. There were also the Ten Commandments, but they did not seem like rules because they were either too obvious to mention (like loving God and honoring my Mother and Father) or irrelevant in my life (like coveting my neighbor's wife or his ox or his ass).

In spite of this profusion of rules I realize now that fundamentally I was taught that rules as made or interpreted by men need not be respected at all costs. The basic lesson drilled into me was that I should try to behave as Jesus did or as He would have wanted me to behave. What He wanted was often not very clear, but at least two things were certain. One was that Jesus himself did not obey rules just because they were proclaimed by human authority, and the other was that He insisted God's kingdom should come "on earth as it is in heaven." I knew, therefore, I should not accept life as it is but should act so that I might help change things in preparation for God's earthly kingdom.

In school I was given the same feeling. History should not bind me. On the contrary, I was shown examples of great men who had swept aside man-made compartments and rules in order to bring about progress. My own ancestors offered good models: they had rejected the restraints of the old world and had come to the wilderness to build a better world. It took only courage to reject the accretion of the ages. In English class I learned by heart:

> Let the Past bury its dead!
> Act, act in the living present!
> Heart within, and God o'erhead!

So I grew up, as the psychologist's child will probably grow up, to reject the idea of a compartmentalized world into which I was born and which I must accept. Life was a continuum, and I was free to move within it to accomplish what I thought right. Between the extreme of the Methodist parsonage in 1910 and that of a psychologist's home in 1960 lies a basic attitude shared by most American families. Man should not be hampered by previously set boundaries.

The differences between the way a French child and an American child are brought up are greatly oversimplified here, but they point to significant generalizations about socialization in France and the United States. The French child learns that life has been compartmentalized by man and that the limits of each compartment must be recognized and respected. The American child learns that life is a boundless experience. The Frenchman recognizes that rules are a convenience, but that they are man-made and therefore artificial. The American believes he has discovered his rules for himself and that they reflect the essential structure of reality. For the Frenchman, reality is dual: there is the official reality of man-made rules, but it is only a façade concealing a deeper, more mysterious reality which may be felt by the individual in moments of introspection or revealed by art and religion. For the American, reality is a unity, and any apparent discrepancy between the ideal and the actuality is essentially immoral.

These two conceptions of reality and of the individual's proper relation to it inevitably help determine the reaction of French and American children to the problems of adolescence. Not only do the problems have a different significance to them, but the solutions they seek are different.

In France the vague but compelling stir of sexual development creates greater wonder and anxiety than it does in America. To a child who has been carefully trained to compartmentalize life rationally, the tempestuous quality of pubescence is threatening. It raises a doubt as to whether these feelings may not become so intense that they cannot be controlled within the framework provided by society. This anxiety is frequently expressed in French literature and films in a way the French find moving but Americans often find frankly ludicrous.

Recently I showed the French movie short, *Les Mistons*, to a group of college students which included both French and Americans. The movie portrays the panic of a gang of boys when they suddenly become conscious of the sexual development of a young girl. The girl and the young man she loves become the enemies of the gang, who react hysterically to the new desires they feel. The boys spy on the couple, torment them, mock them, even stone them. The characters in the movie do not speak, but a poetic commentary describes the situation as it develops. The European members of the audience found the movie a valid artistic expression of a situation that seemed true for them. The Americans found the whole thing ridiculous; the very conception of the movie seemed a joke.

In learning how to cope with his sexual urge, the French child, however, is helped by his training in recognizing proper limits of behavior. He may be more upset than the American child by the feelings that accompany puberty, but he is better equipped to handle those feelings. From early childhood he has been taught the necessity of controlling his impulses, of not expressing them freely. With puberty he is confronted with the most violent urge of all. The urge may frighten him, but he is at least used to exerting self-control.

The American child has had less preparation. Rather, he has been encouraged to express his feelings freely. Now suddenly the attitude of his elders is reversed. He is expected to exert self-discipline and not act out this new feeling at the very point in life when it is most difficult to do so. The injunction that he must restrain his sexual impulse is all the harder to accept, since at the same time the American adolescent learns another hard lesson—that there is a wide gap between the expressed standards of sexual behavior and actual practice. At church and school and home both boys and girls are told that it is wrong to express themselves sexually outside of wedlock. But walking home from school, in locker rooms, at the corner drugstore, from newspaper stories, from movies and magazines and TV, they learn that the ideal standards are false. The American adolescent must choose between observing the standards and feeling frustrated and cheated, or violating the standards, feeling guilt, and risking social sanction.

It is true that there is also a double standard in France. However, this double standard of morality comes as no surprise to a French child, since from the time he was very young he has been made aware of the duality of existence. Furthermore, the French tend on the whole to feel that man must accept the limitations of nature and not try vainly to deny them. Since sex is an integral force in nature it is better to accept and to pattern it rather than uselessly repudiating it or even dangerously distorting it.

A traditional means has evolved in France for the indoctrination of young people in the expression of their sexual feelings. The adolescent boy receives his experience and training from an older woman and then in turn initiates the girl—ideally, of course, his virgin wife—in the art he has learned. French literature and movies offer examples of this, but whether this literary expression mirrors actual behavior is a question which cannot be answered. Still, from what evidence we have, it does seem that the situation so frequently portrayed in the novels of Colette, for instance, bears some relation to actual practice.

To the American boy the very idea of sexual relations with a woman old enough to be his mother seems monstrous. He learns sexual techniques just as he is taught to learn all things—by venturing out, fumbling, experimenting, seeking advice from his peers or from any other source he can find. Eventually, for better or for worse, he evolves his own system. Today in the United States we seem to feel that sexual initiation should come naturally as the result of a burst of romantic passion. We have little information, but it appears that most American boys today receive their sexual initiation on the back seat of a car with girls roughly their own age, who have had their initiation in the same situation with an older adolescent. The romantic situation is enhanced by the beer in a "six-pack" the couple have drunk. This is not a very satisfactory experience for either one of the couple, and it has been suggested that the whole clumsy operation may help account for the feeling of inadequacy shared by many American adults.

The American reluctance to set limits creates another sexual problem for adolescents that is less troublesome for the French— that of determining a feeling of sexual identity. Even when they are very small, French children are rather strictly segregated by sex. Boys and girls are dressed differently, and are given different sorts of toys. They are treated differently and are expected to behave differently. From the age of five or six they are separated in school, even in the public schools. A strong feeling develops of belonging to *les garçons* or *les filles,* and each individual is left no doubt as to his affiliation.

For Americans, on the other hand, it seems wrong to deny a girl or a boy what is accorded to members of the other sex. Consequently, boys and girls are treated essentially in the same manner. As little children they are dressed alike, they may play with the same toys if they choose, and they are treated with studied impartiality. At school they are in class together and compete in exactly the same skills. At home they see the responsibilities of the parents divided by their mother and father on the basis of convenience and with little emphasis on which is properly a woman's role and which is properly a man's. Yet when boys and girls become adolescents, they are told that they are fundamentally different and should feel fundamentally different—or else they risk the dreadful fate of being labeled sexual deviants. With relatively little support, American youth is expected to achieve the proper feeling of sexual identity. The normal bisexuality of humans and the normal homosexual experimentation in our culture add to the confusion. The result is that adolescents feel an urgent need to prove to themselves and to others that psychologically they belong to the class with which they are anatomically grouped. As a result, girls go to extremes to make themselves appear feminine, and boys go to stupid and dangerous lengths to avoid being called "chicken." Homosexuality exists in France, but there is less confusion and resultant anxiety over sexual identity among normal French adolescents than among normal Americans.

By the time a child has reached adolescence he has formed a more or less precise image of what he and his culture expect of him as an adult. One of the principal problems for an adolescent then is to conduct his life so that he has the feeling he is achieving this ideal image of himself.

The average French child (and his parents even more than he) has a clear idea of the limits within which his ambition may be fulfilled. He knows to what social and professional class he belongs. There is no doubt about his family's traditional political, religious, and even aesthetic ideals, and he has been placed by both family and teachers in a well-defined intellectual category. Each of these classifications implies certain limitations and expectations so far as the child's future is concerned. For the normal French child, then, this clear definition of expectations makes the problem of fulfilling his ideal self-image relatively simple. He has only to accept and to live up to what is expected of him.

It is true that many more possibilities are open to French children, especially to gifted children of the farming and working class, than those of which they are aware. There is a tendency in France to assume that one's position in the social structure is fixed. Just as Americans assume there is more mobility in their society than there is, the French assume that there is less chance for change than there actually is. However, this misconception makes it easier for the French adolescent to accept the image he has of himself as an adult.

Of course, not all children can accept or live up to expected limits so readily. The most unhappy French adolescents are those for whom the limits are impossible. The less gifted child of an ambitious middle-class family who tries again and again to pass the *baccalauréat* examination, the independent child who refuses to be harnessed to fixed goals, the imaginative child who has elaborated an ideal quite different from that which society expects of him— such children have a hard time of it during adolescence. These exceptions are not numerous, however. They seem more common to us than they are in fact, because they offer the poignant cases from which novels and movies are made, and we form our idea of French life in a large part through such media. As a matter of fact, most French adolescents do accept the limitations laid down by society and do try to fulfil their ambitions within them.

Acceptance is made easier by a series of escape hatches which French society provides so that the individual does not feel himself annihilated by his acceptance of limitations. The French child has learned that if he accepts social constraints he will be left to himself. There is a tendency, therefore, for the French adolescent to escape into himself, to live in his private domain of thoughts, emotions, and fantasies. Even though one of the most common feelings may be hatred for *les autres*, the people around him who force him to conform superficially, so long as the adolescent's retaliation is confined to his feelings or even to verbal or artistic expression, society imposes no sanction.

Since children have seen that social regulations are in no way sacred but are only a practical means invented by man to permit human beings to live together, and since these regulations are so often frustrating, another traditional means of escape is for adolescents to form cliques and work outside the accepted social bounds to attain a common goal. The presence of this clandestine association, which Jesse Pitts has called the "delinquent peer group," is felt wherever children need to get things done in spite of rules. Americans visiting French schools are shocked by the hidden power exercised by an apparently well-disciplined class over their teacher. An effective *chahut* carried out by the conspirators may even ruin a teacher's professional career. A child who runs afoul of the con-

spiratorial system may be punished by the group in a way that
seems shockingly cruel. The clique permits adolescents to accept
social limitations and at the same time reject them. Furthermore,
participation in a clique brings one of the cherished experiences of
life—warmth of mutual association in a secret, illicit endeavor.

Among older adolescent boys this association is encouraged by
society. In a sedate, conservative village in western France I have
seen the *conscrits*, the group of nineteen-year-old boys who will go
off next year for their military service, spend every Saturday night
over a period of months getting drunk on wine furnished by their
parents. One evening six of them were said to have drunk thirty
bottles of wine while they waited for the rest of their comrades to
arrive. This seems impossible and probably was, but at least it is an
indication of the massive scale of these binges. They are not only
tolerated but encouraged by adults as long as the adolescents do
not jeopardize their future.

Generally speaking, then, society helps make the severe limitations it places on children more bearable. To fulfil his self-image,
the French adolescent must learn to fit into the limits prescribed by
society, and having accepted these boundaries, he utilizes the means
available to express his individuality outside these limits.

The American adolescent is in a quite different predicament as
he strives to achieve the ideal image he has formed of himself. His
difficulty lies not in living up to expectations but in discovering what
they really are. The only system of rules he has been taught is a
Sunday-school sort of code, and as he grows up he learns little by
little that it is not the code by which people actually live. The real
code exists, but no one defines it openly.

Confronted by the fact that this double standard exists, American adults beat their breasts and admit their sins. Still they insist
that the ideal code is the right one. Failure to live up to it is
attributed only to the weakness of human beings who hopefully
merit forgiveness when they confess their sins and show their good
intentions. The adults' need to believe that everything will come out
for the best in the long run is satisfied by placing the responsibility
on the adolescents to make the ideal code function as it should.
"Our generation has gotten the country into trouble, and we want
you to get us out," Senator Barry Goldwater tells an audience of
Young Republicans,[1] just as every adult speaker has told every
adolescent audience in which I have been present since I was a
child. Adolescents are not told how to do a better job, however.

When they ask for advice, they are merely given further indoctrination in the ideal code. Middle-aged professional adolescents continue to insist at Sunday-evening meetings, summer conferences, in discussion groups and recreational organizations that the ideal is attainable. Dozens of books are written for teen-agers to be used in these meetings.

A typical book for adolescent instruction is Pat Boone's *Twixt Twelve and Twenty*,[2] in which the blurb says "Pat talks intimately to teen-agers about all the problems and joys of the exciting years." There are chapters on life such as "A Great Adventure," on "The Happy Home Corporation," "Habit Weaving," "God is Real," and "Dreams Do Come True." My informants tell me this book is rarely bought by adolescents themselves, but is purchased in large quantities by church organizations for use in young people's groups.

The reaction of most adolescents to the duplicity of adults is an outcry against sham and an increased emphasis on the value of sincerity. Newsstand dealers will tell you that there are four kinds of publications bought by adolescents themselves: comics, movie and confession magazines, technical magazines (cars, mechanics, sports, etc.), and above all *Mad* and its imitators. *Mad* is a specific antidote for Pat Boone. It shows that all the virtues which are extolled are in reality only a screen for vice. *Mad* delights adolescents by turning the official American value system upside-down.

In a recent number of *Mad*[3] from the collection accumulated by my children, the first article suggests a whole series of new textbooks to teach children life as it really is: "Today's children are developing into clods because old-fashioned textbooks still in use fail to hold their interest, fail to reflect life as it is lived now, and fail to prepare kids for what they face in the years ahead. To remedy this situation, we recommend that schools immediately junk their outdated texts and replace them with *Mad's* Modernized Elementary School Textbooks."

The proposed third-grade arithmetic is composed of problems like the following: "1 Elvis plus 12,000,000 Teenagers at 16 shrieks per Teenager equals? Answer: $3,000,000 a year for Elvis!" The new geography manual would not have the traditional figure of Atlas on the cover but two Atlas musclemen, each supporting a half of the world. One is labeled "Good Guys" and the other, "Bad Guys." Children would be introduced to the *First Principles of American Civics* by a cover illustration showing a bum being paid to vote by a politician wearing a campaign button reading, "I like

Me!" The following chart of real municipal government in the United States is shown on a sample page:

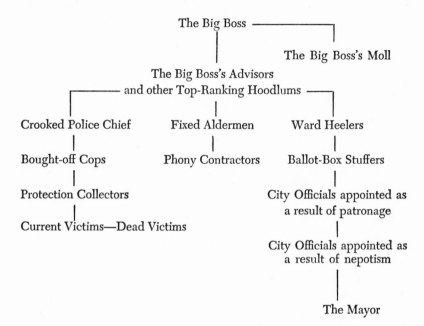

The Big Boss ─────────┐

                        The Big Boss's Moll

The Big Boss's Advisors
and other Top-Ranking Hoodlums

Crooked Police Chief   Fixed Aldermen   Ward Heelers

Bought-off Cops   Phony Contractors   Ballot-Box Stuffers

Protection Collectors   City Officials appointed as
                              a result of patronage

Current Victims—Dead Victims   City Officials appointed as
                              a result of nepotism

                                The Mayor

In *Mad* the adolescent finds a confirmation of his discovery that adults have furnished him a false blueprint of life. Government, school, family life, advertising, TV, movies, business—even *Mad* itself!—are exposed. *Mad's* symbol, the insipidly smiling Alfred E. Neuman, who maintains his ghoulishly cheerful expression while the most appalling things go on around him, stands for American culture itself as the adolescent experiences it. This idiotically smiling figure implies that all is for the best in the best of all possible worlds, in spite of overwhelming evidence to the contrary. Alfred E. Neuman is the American *Candide*, but with one difference: he never mentions God and religion. Even *Mad* could not get away with that.

The anguish an adolescent may feel when faced by the lack of a realistic moral code is portrayed in the James Dean movie, *Rebel Without a Cause*, a movie that draws a large audience of adolescents whenever it is shown. In the movie young Jimmy pleads again and again with his father and mother to tell him what he ought to do. He wants to know the rules. His parents shrink before his questions.

His mother's only solution is to move the family to a different city every time Jimmy gets into trouble, in the hope that in a new atmosphere Jimmy will "behave." His father will reply only that "the important thing is to understand, just to understand." Jimmy is left to work out his problems for himself, in spite of his plea: "Tell me what to do! Please, Dad, just tell me what to do!"

It is painful for American fathers to witness this scene, for we as a group are committed to Jimmy's father's position. We believe that we need only give our children loving understanding, the Sunday-school rules, and a *Reader's Digest* philosophy, and that they will work it out for themselves. They usually do, but it is a hard assignment we give them, and one which the French adolescent is largely spared.

As French and American adolescents face the problems of realizing their self-image, the difficulty for the French is to live up to or contain themselves within the limits society had laid down for them. The difficulty for American adolescents is to discover behind the façade of the ideal system the real limits established by society.

The normal desire of young people in France and the United States to establish homes and families of their own poses another question. As we have seen, the contrasting attitudes toward limits have a direct bearing on the ways in which adolescents in the two countries face the problems of learning to live with their physically maturing bodies and to fulfil their image of themselves. This same influence is felt in the way the problem of "hiving off" is resolved in the two cultures.

The French adolescent is left in no doubt when questions arise concerning his role in the family. "So long as you live in this house, you'll do what we tell you to do!" says the French father in the normally regulated household. Although the mother may soften the harshness of this decree when she deals individually with the children, traditionally she backs up her husband. The child, the adolescent, and even the young adult child still living at home must accept what is at best the benevolent despotism of the father, although as we have seen, the father limits his authority to the realm of the action and usually does not try to control the ideas and feelings of his children. With his experience of clandestine action, an adolescent may be able with the connivance of a sibling, sometimes the mother (and sometimes even the father himself!) to carry out an action that may be formally forbidden. Nevertheless, there is no doubt where the seat of authority lies and what the limits are. When the chips are down, it is the father with the advice and consent of the mother who makes the decisions which regulate the activities of the ménage.

The dependence of the French adolescent on the family is increased by his lack of personal financial resources. The money a child earns is not his own but is given to the parents for the family treasury. This is true not only of young adolescents but often even of young, unmarried adults living at home. There are exceptions. A young middle-class boy or girl may work in the grape harvest in the country near the place where his family spends vacations and save the money to buy a scooter, but even in this case his keeping the money and buying the scooter must be approved by the parents.

Of course, a French father is obligated to give his child sufficient money for the leisure activities of his particular social group. A father who is too stingy is frowned on. Even worse, however, is the father who indulges his child too much. Recently, in the murder trial of Georges Rapin, "Monsieur Bill," who was eventually executed for killing a young woman in a singularly brutal manner, it was not the murderer himself who was most severely censured. The newspapers and the public vented their anger against the father. He had spoiled his son by giving him far too much spending money. He had even bought the young man a bar when he refused to take the respectable sort of job that might have been expected of an upper middle-class engineer's son. When the boy was being taken to the death cell after the trial, people followed the father out of the court and jeered him along the street. *Match* and other popular magazines carried photographs of the incident and pointed out in editorials that this was a good lesson for parents who do not teach their children to behave reasonably.

The older a French adolescent is, the more he tires of paternal authority and the ever repeated phrase: "So long as you live in this house. . . ." The effect is to strengthen his determination to become independent. As soon as he finishes his military service and professional training he tries to make a clean break and establish his own ménage. In France, as in the United States, the age for marrying has been pushed back toward adolescence, and like American parents French parents complain that children are marrying too young. The reasons for earlier marriage are different. In the United States, marriage is a romantic refuge from the bewilderment of life; in France it is a refuge from the severe restrictions of life.

In American culture a child is taught very young to fend for himself, to make his own decisions, to stand on his own feet. A good illustration of this is in the movie, *Four Families,* produced by the Canadian Film Board under the general supervision of Margaret Mead. Of the four babies shown in the film, the Anglo-Canadian baby was, of course, the most spoiled by modern conveniences. However, in some ways she had a harder time than the Indian, Japanese, and French babies because she was expected to care for herself in a way the other babies were not. During the ten-minute episode, the Canadian baby was constantly urged to venture out and take care of herself. She got a hard bump on the nose when she was encouraged by her father to take part in the rough play of her older brothers. When she cried, she was merely told, "There, there, it'll be all right." In the bath she was encouraged by her mother to fight for the possession of a washrag held by the mother herself. At the table she was expected to feed herself. In contrast to the other babies, who were fed again and lulled to sleep by their mothers at bedtime, she was put in her crib with a bottle and left alone to go to sleep.

As children in America grow older, we continue to encourage their independence. We encourage them to make money, to start their own bank account. They run errands, baby-sit, carry newspapers, sell lemonade—in fact, they often surprise their parents with the jobs they turn up on their own. They use their earnings as they want. "They are at a learning stage. I cannot think of any better way to teach them than to let them do it," says Gwen Bymers in support of this training in independence.[4] American children are also encouraged to make their own social ties outside the family. Instead of being limited, as French children are, to a group of friends known to the family circle, they venture out and make their own friends. American parents even utilize their children's friendships to make the acquaintance of the parents of their children's friends. At a very early age children here come to feel at home beyond the limits of the household. The fact that their household is open freely to outsiders further blurs any clear conception as to what home is exactly. There is not the wide, clearly defined gap between home and society that there is in France.

Within the American family there is seldom any insistence that the family has a way of doing things which a child must accept so long as he lives under the family roof. More often the child's argument, "But nobody else's family . . ." sets the standard of behavior. In deciding family affairs the child is given an equal vote with the parents, whose attitude is characteristically expressed by the phrase, "Let's work this out together." When a problem arises in which the adolescent's and the parents' judgments are in conflict, the outcome is ideally a compromise by which the child may get what he wants at the same time as the worries of the parents are allayed. There is no real tyranny in the American home unless it be that exercised by the adolescent who shames his parents for their unacceptable social behavior.

Far from being the autocrat who lays down the law in the family, the American father tries to become his children's friend so that they will want to turn to him for advice in solving problems that face them. Sometimes this friendship comes about naturally, sometimes it is forced. The accepted social code thoroughly approves this kind of comradeship. A caricature of the situation is described by Dobie Gillis, the "pint-sized Don Juan," adolescent hero of Max Shulman's *I was a Teen-Age Dwarf:*[5]

I'm a little embarrassed to tell you about our Palship Walks, but I guess I'd better. It's one of my mother's kooky ideas, which Pa and I fought against like a couple of madmen, but it wasn't any use at all because when Ma gets an idea in her head you can't knock it out with an elephant gun. . . . Ma got on Pa's back a few years ago about him not spending enough time with me. "Herbert," she screamed, "a man ought to be pals with his son. Why don't you take Dobie for walks on Saturday morning to talk to him about nature and engines and like that?" Well, Pa and I both started yelling like maniacs because we didn't want to go for a walk on Saturday mornings. What I like to do on Saturday morning is crack my knuckles. What Pa likes to do is stay in the sack. It is the one morning he doesn't have office hours. But Ma just ignored us and put on our jackets and pushed us out of the door.

So Pa and I stumbled around for a while, and it was pretty grim. At first he tried to talk to me about nature and engines, but that didn't work too well because I kept thinking about cracking my knuckles and he kept thinking about the sack. Finally we sat down against a big oak tree on a point of land overlooking the ocean and moped till it was lunch time and we could go home.

After that we didn't make any attempts at conversation on our Palship Walks. We just high-tailed it out to the oak tree where Pa had stashed an air mattress in a hollow limb and I had stashed a copy of *Lolita*. Pa blew up the mattress and corked off for a couple of hours while I read the book and then, both refreshed, we went home where Ma beamed at us and kissed us and gave us a special treat for lunch in honor of our palship.

And, of course, since "mothers know best," the father and son by sharing this exile did come to feel close enough for the boy to ask a question once, and the father gave him a straight answer. But the father had had to earn the right to help his son make a decision. The boy's business was the boy's business, after all, and the father had no right to interfere without being asked.

Ironically, although the American adolescent is encouraged to be independent, although he is free to make his own decisions, he is in no hurry to leave home. Life at home is comfortable. Even though he may marry young, he feels no need to make a clean break. Unless the young couple move so far away that they are of necessity separated from their parents' homes, they still linger about their old haunts and depend on their parents to help them as they always did. They drop in to have a wedge of Mom's cake or a drink of Dad's whiskey, or to get help in making draperies, or in giving a party. Eventually the parents become admirable baby-sitters. They cost nothing, they can be counted on to stay with the baby until all hours of the morning, and without doubt they know best what to do in case of an emergency. Eventually, it is the parents of the adolescents, not the adolescents themselves, who long for a clean break between the old home and the new home.

How different this is from the situation of the French couple who have had enough of the discipline of the parents' home and are eager to make the break themselves! In France it is the young couple which treasures its independence, maintains proper relations with the older generation but still insists on handling its affairs alone. The unhappy young French couple is the one that cannot find adequate housing and is obliged to live under the parental roof.

In the light of early training, how is this difference to be interpreted? Should the emphasis be on the confidence the young French have acquired from learning early and well what the limits are? Or should the emphasis be on the frustration these limits have created?

The same question must be raised concerning the American adolescent's reaction. Has his lack of feeling for what is and what is not made him unsure and tired? Does breaking away from his parents' home imply too great an effort, too much more searching? Or has his family experience made him so comfortable with his parents that he does not want to leave them?

Having begun this essay with broad generalizations concerning the difference between French and American culture and the impact of these differences on the lives of adolescents in the two cultures, we come now to the point where we no longer generalize so confidently but raise questions about what might be the truth. Who knows for sure the exact effect of child training? Doubts concerning our generalizations are strengthened by other considerations. Although we intended to speak of all adolescents, it is obvious that what has been said mainly concerns boys. Do these same generalizations apply to girls? Certainly not without some change.

Recent conversations with French parents raise another question. They complain now that their children have no respect for them, criticize them, insist on establishing their own rules. Is my picture of French family life outmoded? It would be so much simpler to speak of differences between American adolescence and adolescence in the Nuer tribes of the upper reaches of the Nile River. Then we could point to striking, definite contrasts about which there could be no doubt. Comparatively, the differences between French and Americans seem so slight. And yet they exist. No one who knows French and American culture will deny that there are substantial differences. How can we define them?

Recently a French colleague said that from his point of view French and American parents are today in the same predicament. He knows "how uncontrolled, how lacking in respect for tradition" American youth is. But times have changed in France. Children no longer accept the limits set by their parents. This French father says he could not get away with saying, "So long as you live in this house you will do as I say!" His children would only laugh and still do as they please.

On the other hand, there is more talk in the United States today about the importance of giving children a feeling for limits. In a recent article,[6] Ann Landers, human-relations counselor and newspaper columnist, suggests that American parents have allowed their children far too much freedom for their own good. Rules must be set and accepted by young people whether they seem to like it or not. This sounds like a step toward the French concept of discipline.

However, Miss Landers and my French friend are still far apart in their attitudes. The French adolescent children are not revolting against the very idea of limits. They are revolting against the particular set of rules which their father insists they accept. This sort of revolt is very much within the French tradition and in accordance with the idea that the young Frenchman insists on differentiating himself from his parents.

On the other hand, Miss Landers wants parents to assert rules as little as possible. She does not praise limits for their own sake:[7]

> All parents should allow their teen-agers to make a great many decisions for themselves. The vital question is where to draw the line. But wherever the line is drawn, remember that even a foolish decision can be a useful one if the teen-ager learns a lesson from the mistake. Just take care that you don't bail your youngster out of every embarrassing situation. . . . Your major job as a parent is to equip your child to lead an independent, productive, useful life. Live with your child, not for him. For the most part, let him take his own jumps but don't let him jump off a cliff to learn first-hand what's at the bottom. Be firm but be fair. Respect him and his rights and you won't have to worry about his respect for you.

When my French colleague is no longer shocked at his children's revolt, when he insists that they learn by making mistakes, when Miss Landers insists on the parents trying to pattern their children's lives, not merely preventing them from jumping over the cliff, then I shall agree that American and French culture have indeed become very much alike. Until then I shall continue to believe that this difference in attitude toward limits is a major factor in differentiating not only adolescent behavior but also many other aspects of French and American culture.

REFERENCES

1. *The Harvard Crimson*, 3 May 1961, p. 1.

2. Pat Boone, *Twixt Twelve and Twenty*. New York: Prentice-Hall, Inc., 1958; and Dell Publishing Company, 1960.

3. *Mad*, No. 63, June 1961.

4. Gwen Bymers, in *The New York Times*, 8 May 1961, p. 47.

5. Max Shulman, *I Was a Teen-Age Dwarf* (New York: Bantam Books, 1960), pp. 3-4.

6. Ann Landers, "Straight Talk on Sex and Growing Up," *Life*, 18 August 1961.

7. *Ibid.*, pp. 84, 88.

GEORGE SHERMAN

Soviet Youth: Myth and Reality

"LET'S SUPPOSE a man lives seventy-five years. For the first twenty-five he should study, for the second twenty-five he should travel the world, and for the last twenty-five he should rest, reflect and write—preparing for death."

Yuri, a twenty-nine-year-old Soviet geologist, was philosophizing. We were relaxing together on the fashionable hotel veranda at Sochi, the popular Black Sea resort, sipping Soviet champagne and savoring summer breezes from the dark water beyond.

Soviet propagandists would be aghast at Yuri's thoughts. In their mythology he is no ordinary geologist: he is one of the most dedicated of "the young builders of Communism." For the past four years Yuri has forsaken his native Moscow to drill for precious minerals beyond the Arctic Circle, near Vorkuta in European Siberia. For ten months of the year he lives on this North Soviet "New Frontier" in a small dormitory room of a hotel-hut. Vodka and hard work are daily sustenance on the frozen tundra. Soviet newspapers sing hymns to self-sacrificing young people like Yuri, who are supposed to be driven by love of country and devotion to Communist duty.

In Yuri's case, at least, the incentives are more prosaic. First, he is paid an inviting salary of 4000 roubles monthly. The average Soviet wage is 800. "Here in the South or near Moscow I could receive no more than 1500 a month," he said unashamedly. Second, Yuri receives two months' paid vacation every year. Normal vacations are three weeks to a month. Third, although Yuri has volunteered for his particular assignment, he pointed out that every Soviet graduate "pays" for his free education by working at least three years where the state employment office sends him. According to press reports, many of the educated young people hatch incredible schemes to evade work assignments outside the large metropolitan centers.[1] Finally, Yuri is not a Communist, nor does he desire to be one. "Membership in my professional union (*profsoyuz*) is enough for me," he said. Politics do not interest him, he made clear, and he became noticeably uncomfortable when they were mentioned. He dismissed a question about Stalin's forced-labor camp in Vorkuta with the simple explanation that all political prisoners had departed before he arrived. "Anyway, I have little contact with the local population," he said, and dropped the subject.

On the other hand, Yuri's curiosity about the outside world was insatiable. He had arranged to have his vacation in Moscow during the International Youth Festival in 1957. Rome was first on the list of foreign pilgrimages he planned. "Contacts, contacts and more contacts—that is the only way we Russians will learn to understand you foreigners," he exclaimed. Next year he would quit his job and return to Moscow to take a postgraduate degree. That would give him both knowledge and status (he already had money) to further ambitious travel plans. "The way things are going, I may visit North and South America in five or six years," Yuri prophesied. Who could contradict him? Such optimism is endemic in the Soviet era of "rising expectations."

To meet young men and women like Yuri in the Soviet Union today is nothing unusual. To meet young people who are *not* like Yuri is also nothing unusual. One of the great changes in the post-Stalin period is the greater scope the enquiring outsider finds for probing the human variation behind the monolithic façade once presented to the world. The younger Soviet generation are the first complete products of Stalin's system. Anyone younger than his early thirties has had his whole life molded by a complicated combination of organized persuasion and police coercion. From the day a child can first understand the spoken word, he is gradually taught to subordinate his own desires to the demands of totalitarian society.

Against this background, the steady clamor in the press about "bourgeois hangovers" in the younger generation has a hollow ring. The West cannot be blamed for blemishes in the new socialist society. For over a decade after World War II the development of Soviet youth proceeded in almost complete isolation. Whereas their parents bore the brunt of forced industrialization and the forced collectivization of agriculture, this new generation has been nurtured on the havoc of war, schooled in the particular horrors of Stalin's reconstruction, and graduated into Khrushchev's "peaceful coexistence."

From this environment the new "Soviet man" is supposed to emerge. In literature, on wall posters, in newspapers, over the radio and television, he looms out larger than life. As an ardent member of the Young Communist League (the Komsomol),[2] he willingly devotes his leisure to "socially useful" work, he conforms morally and politically, he disciplines his innermost life to the will of Communist authority.

Mr. Khrushchev's drive toward "normalcy" has somewhat blurred that image. The model is still there, but it seems to have lost some vitality. Less arbitrary police controls, limited prosperity, and a general relaxation seem to have had their greatest effect on the young. They are less timid than their elders (who remember Stalin's purges) in exploring the bounds of freedom. In the process, many of the would-be robots are displaying the same virtues and weaknesses of youth the world over.

My purpose is not to analyze how far this virus of experimentation has spread among young Soviet citizens, or to estimate its ultimate impact on the society. Soviet society is still a very closed one, despite increased contacts with the outside world. The old myths are still religiously (although more gently) fostered from the top down. Numerous insights are now possible into underlying reality, but they are still only isolated fragments of unverified truth.

My purpose is rather to describe types of Soviet young people and their problems in the post-Stalin environment. Some have adjusted to that environment, some have not. My observations are based primarily on intermittent experiences with such people as Yuri over a five-year period (1955-1959). Wherever possible, they are supplemented by increasingly frank revelations on the part of the Soviet press and Mr. Khrushchev himself.[3] My main focus is on the large industrialized cities, whose better standard of living, cultural life, educational institutions, and factories attract congregations of young men and women. In these centers the persistent tussle between old myths and new realities is more readily observable.

The young people who have attracted most public attention at home and abroad are the *stilyagi*, the "style-chasers," the rough Soviet equivalent of the British "teddy boys" or the American "drugstore cowboys." In the beginning the *stilyagi* were not necessarily juvenile delinquents. They were boys and girls who appeared in public in "Tarzan" haircuts, bright American-style shirts and too narrow trousers or skirts—teenagers and beyond who called each other by American nicknames, illicitly recorded American jazz, and made primitive attempts at rock n' roll. In the end, however, the *stilyagi* have become a catch-all label in popular parlance for "antisocial" conduct—the "hooligans" who create drunken brawls, the black marketeers or "center boys" who trade counterfeit ikons for foreign tourists' clothes around the central hotels, the "gilded youth" who use their parents' influence to evade social duties and responsibilities.

A delving press has uncovered the spread of special slang among youth—labeled a "stilyaga-ism of speech." Young band musicians have begun to refer to their engagements as "playing at a funeral." One writer overheard the following conversation between students: "Well, let's fade. I still have to hit the hay; there's going to be a big shindig at a pal's shack."

The writer was even more dismayed at finding the "linguistic nihilism" spreading from students to workers. Instead of "Let's eat!" they are beginning to say, "Let's feed," or "Let's chop." Instead of buying something, the young workers now "grab" or "tear it off." The word "mug" is now used instead of "face," and the television set (*televizor*) has become the "telik."[4]

The line between innocent innovation and criminal delinquency is hard to draw in the Soviet Union. Any action not officially inspired and controlled is potentially dangerous to the regimented society. A kind of Victorian puritanism, inflexible and humorless, dominates the scene. Established authority makes full use of it to stamp hard on all overt signs of nonconformity. A vicious campaign in the press has sent the most extreme fads underground. Informal "comrades' courts" have been set up in factory and office to criticize and ostracize minor offenders—the girl who wears too much lipstick or the worker who arrives at work with vodka on his breath.

In Ivanovo, a textile town 300 miles northeast of Moscow, the sprawling Park of Culture features a row of life-sized posters ridiculing young culprits most recently apprehended by the People's Militia. These public caricatures change every two weeks. A girl worker, with vodka bottle at her lips, lounges lazily on the sunny beach; Marina has spent two weeks in jail, the doggerel underneath explains, "because the sun made her insides too warm." Another caricature shows feline Alexandra, a "beautiful cat," preening herself before the mirror, "because she would rather have a man than work."[5]

The protection of standard culture for the masses has collided head-on with the drive for sophistication among educated and quite respectable young people. On the one hand, they are exposed to an increasing number of fashion shows, articles, and books on refined manners, and to more limited numbers of foreign cultural imports such as British and French film festivals, a Polish exhibition of modern art, a Czechoslovak glass exhibit, and several American exhibitions. On the other hand, the new fashions which spread from the capital to the provinces fall victim to organized Komsomol scorn.

Letters and articles in the Moscow press complain that roving bands of "Komsomol police" have hunted down and molested vacationing young men in bright shirts and young women in slacks on the streets of fashionable Sochi.[6] Stylish girls have had their hair chopped off with "sheer violence." Two young girls from the distant province of Amur in Siberia complain that they were treated "like *stilyagi*" in their village because they wore one-piece fitted dresses. Their Komsomol Committee told them: "Dress so that you will not be different from others!"[7] *Komsomolskaya Pravda*, the chief youth paper in the Soviet Union, has had to reassure Kiev residents that the "music patrols" set up by the Kiev Komsomol do not have the power to prohibit the playing of "good jazz."[8]

The effect of the excesses has been a slow swing of the social pendulum toward some kind of compromise over developing tastes. Official attacks on intolerance bring with them demands for an "ethic of mutual respect."[9] In practice, that seems to mean greater individual freedom in private or semiprivate, while paying lip-service in public to slowly changing social conventions. Public dance halls in Moscow, Stalingrad, and Sochi all still look much the same: young people in open-necked shirts and shapeless dresses move around crowded floors to conventional waltzes, some folk-dance music, and nondescript foxtrots for the "masses." Komsomol police see that no one steps out of line. In the expensive restaurants and hotels, however, an air of relaxed sophistication is becoming more noticeable. In the National Hotel dining room in Moscow or the Gorka Restaurant in Sochi, popping champagne corks punctuate the occasional cha cha cha, rock n' roll, and other improvisations the dance bands have learned from the Voice of America. Toward the end of the evening several well-dressed couples may prove that the ban on "unorthodox dancing" is not uniformly enforced.

Changes are also going on just beneath the surface in the field of art. Although "abstractionism," or "formalism," or "subjectless art" (in fact most experimentation) is officially condemned, some of the younger intelligentsia are tasting the forbidden fruit. A network of "private" Soviet art is spreading in the cities, aided by Mr. Khrushchev's drive for socialist legality. A friend in Leningrad who collects and disperses "modern art" for artist friends said he now had little fear that the police would invade his rooms simply to remove paintings from the wall. Officials of the Artists Union had threatened to inspect the studio of one of his clients, a well-known artist. They suspected he was dabbling in unorthodox art—as indeed he was. The artist (who is also well-known for his war record) bluntly replied he would throw out anyone who entered without permission or a search warrant. He would publicly display what pictures he chose. Any others were his own affair. The artist evidently carried the day, for the threat was dropped, and he is still painting.

The drive against the *stilyagi* has hampered but not destroyed the development of (Soviet-style) "beatniki" among the younger artistic and literary intelligentsia. Criticism has made them more discreet, less flamboyant. Neither by past training nor present desire do they reject society. Not even the most radical would follow American beatniks in debunking the central tenet of Soviet life: the sacredness of work. As a young Soviet writer put it to me, anyone who does not at least pretend to work is soon investigated and chastised by his local "block committee."

The tendency of Soviet "beatniki" is to emulate what they consider the Left-Bank bohemianism of Paris. It is a faint whisper of a similar movement among young East European intellectuals, particularly in Poland, to make ultrasophistication their mark of separateness from "proletarian" society. In the semiprivacy of their artists and writers clubs, or in their homes, they may don the long cigaret holder, dark glasses, bright orange lipstick, or tight skirt. Perhaps the closest thing to public spontaneity comes in the groups of young people who gather on summer evenings in Moscow to read their poems before the statue of the Soviet poet-hero, Mayakovsky. The stereotyped imagery of socialist realism still predominates, but innovation is more evident. University newspapers and their literary supplements also begin to allow more scope for individual creativity. Take, for instance, the other-worldly quality and quietly religious resignation in the following lines.[10]

The Ikon

    In a dusty, cobwebby attic

Turning its face from the world
An ikon with gilded crown hangs over a pile of torn boots.
For a time moonlight steals in through the cracked roof
Lighting evanescently
The red-haired one, unshaven with grey dust,
Yet still God's haughty image. . . .
Let the mice quietly gnaw each other to death
While he, masked like a pill-box,
Sees and hears nothing in the attic
But waits. He awaits something nonetheless.

The controversy over modes of dress and social behavior is much more than the Soviet version of a universal problem with youth. It reflects a much deeper social conflict in the Soviet Union: the conflict between stifling paternalism and rebellious youth, characterized in less regimented societies as the "conflict between generations." According to Communist mythology, this conflict cannot exist in socialist society. All generations are supposed to be helpmates along the predetermined road to a new heaven on earth. In fact, however, Soviet society today is grappling with a central paradox. Old controls imposed in the name of that heaven are increasingly questioned by the young reapers of half-way prosperity.[11]

The problem can be sensed in the perplexed words of a middle-aged engineer, speaking about his son: "Sometimes I do not understand him," he told me, "he wants everything in the world right away. He thinks too much about 'me,' and not enough about 'us.'" For this man and his generation, serving the "collective" was an imperative of survival. Stalin had been their idol; the suffering and sacrifices of the 1930's and 1940's, their religion.

For the young, this past lives only in stories or dim childhood memories. Their Soviet society is no longer revolutionary, it is established. Relaxation is as evident as is the increase in consumer goods that heralded it. For forty-odd years the Russians have been loudly beating their collective chest about the glories of building socialism. Now it is built. Many doting parents encourage their children to enjoy the youth they lost. People can afford to be less vocal and more confident. They begin to worry less about the socialist image and more about the substance of their own lives.

The virus of easy wealth is most deadly among the *lumpenproletariat*. Many of the real delinquents covered by "stilyaga-ism" are recruited from their ranks. They become the professional speculators who make their living on foreign goods, currency, and innumerable domestic rackets. Their inner bearings differentiate

267

them from the *nouveaux-riches* and even more elite "gilded youth" who lavishly spend their own and their parents' money on scarce luxuries they consider stylish.

By comparison, the young speculators have been raised in working-class slums. They have seen their parents (many of whom were peasants new to city factories) work long years for a single room in a prerevolutionary apartment shared with four other families. One young worker I visited had obtained his privacy by a pasteboard partition around one corner of the room. The whole family had to cook over one gas jet on the communal stove in the communal kitchen of the apartment. Given such cramped quarters and full working hours for each adult—raising children is more a responsibility of the "collective"—family life virtually disappears and the harsh life of the street takes over.[12]

The young speculator begins with a yearning to break out of these surroundings. He is impatient for a better lot, without the hard work which has brought parents and friends small reward. He dismisses the modest improvement of living standards since Stalin's death as inadequate. For him the glamor of Western "easy living" is irresistible. The means for making a quick—if illegal—fortune are close at hand. The advent of more foreign tourists, more Soviet trips abroad, and more private wealth combines with inferior consumer goods and a chaotic distribution system to provide a golden opportunity for black marketeers. Despite official strictures, these speculators make a good living while they fulfill a real economic function.

The price they pay is high. They are not only outside the law; they are totally outside the pale of respectability. And respectability counts for much in Soviet Victorian society. These young people dress well, eat and drink the best food and wines, but they are still *déclassés*. An amateur Soviet sociologist, himself quite respectable, described the type:

> They become empty human beings. They have nothing but their own fine appearance. They set out to push themselves to the top, but they end up belonging nowhere. No good family will have anything to do with these *stilyagi* except for "business."

Superficially, the wayward "gilded youth" have many of these same antisocial habits. They dabble freely in illicit foreign goods, from clothes to books. They are certainly as lawless and immoral. They take out their boredom in wild drinking bouts and parties in their parents' apartments or country *dachas*. They make free use of the family ZIM limousine, which technically belongs to the State ministry or enterprise. In private, however, these reprobates are regarded as the Soviet equivalent of Shakespeare's "Prince Hal." The parents' position in society constantly pulls them toward respectability, while the parents' influence prevents the "adolescent flings" from becoming public scandal. Time and age are supposed to bring the middle- and upper-class delinquents back to the confines of conventional society.

It must be emphasized at this point that the maladjusted young people described above do not mean that the whole rising generation in the Soviet Union are problem children. Quite the contrary. The bulk of Soviet young people emerge as conventionally minded as the Communist apparatus intends. This is particularly true of the young workers. My amateur sociologist friend described them as the "gold of the system." They are hardworking yet submissive. Paternal authority channels their thinking toward ever greater material rewards for increased production. Young "Stakhanovites" and "brigades of Communist labor" are praised at every turn for setting the pace on the production line. Professional Komsomol "students" lead extracurricular political lectures and study groups in the factories after work hours. The factory collective, run by factory committee and trade union, makes sure they are integrated into social clubs and sports programs. Young workers are further encouraged to study in factory trade schools and technicums. The more intelligent and hard-working still have access (although more limited) to the prized professional institutes and even the university.

The result is a political passivity which seems assured so long as the over-all system remains stable. In the words of the sociologist cited above: "These young workers may grumble and protest when they receive 800 roubles bonus at the end of the month instead of the 1000 promised, but they soon tell themselves that 800 is better than the 600 they used to receive. So long as they have enough money and more things to buy, they are happy enough."

The same attitude is roughly true of most institute and university students, although they have scarcely any affinity with the working class. They judge the value of their education according to the status and money it will earn them, not on its intrinsic worth. In this, they are the logical heirs of Stalin's abandonment of egalitarianism in the early 'thirties for more prosaic reward incentives. All the powerful and influential occupations established during that period of the "building of socialism" can now be obtained only through the higher institutes and universities. Furthermore, the number of openings in each category are strictly tied to the needs of the plan. These limitations, plus the social values of a newly industrialized society, have dictated the upgrading of some professions and the downgrading of others. For instance, students specializing in engineering, chemistry, and physics have a higher prestige and will earn higher salaries than those in the humanities (a preparation for teaching), in law or in medicine—careers that are not in such great demand.[13]

The overwhelming emphasis on scientific disciplines is borne out in the educational statistics. In a thorough-going study, Nicholas DeWitt has found that in the sample year 1954, 60 percent of Soviet classes graduating from higher institutes and universities were majors in engineering, physical, and other natural sciences. This figure excludes graduates of scientific pedagogical institutes which train secondary-school teachers. In the field of higher graduate and research degrees, the figure is even higher: 70 percent of all advanced degrees in 1954 were in scientific fields.[14]

While the demands of the newly industrialized economy and the built-in bias of the economic plans undoubtedly promote this popularity of applied and theoretical science, powerful psychic and monetary incentives reinforce the appeal of the career for the young. My friend Yuri, vacationing in southern Sochi from the far North, is an example of the salaries and the opportunities open to young engineers. But any young man—or woman—chosen for research in one of the institutes at the top of the scientific hierarchy is also assured of working with the best equipment, of having every possible professional resource available, and of receiving good housing in the attractive (but crowded) urban centers. Success while young also holds out the promise of higher salaries, higher status, and the greater personal freedom given senior scientists.

Science offers the greatest possible retreat from politics in a system in which everything is political to a greater or lesser degree. As will be discussed shortly, young scientists, like other young people in institutes and offices, are subject to constant political supervision by Party or Komsomol "activists." If they overtly transgress the bounds of political orthodoxy outside their work, scientists suffer the same Party reprimands, loss of career opportunities, or even imprisonment, in extreme cases.

On the other hand, *inside* their work, research scientists (with notable exceptions such as biologists) do not have to spend their lives dodging Party doctrine. The intellect remains relatively un-scarred by the demands of dogmatic truth. Unlike students or young professionals in such politically sensitive disciplines as history, art, and the humanities, scientists have an autonomy free of the day-to-day dictates of shifting ideological interpretation. Of course, this adds to the appeal of scientific studies and detracts from the appeal of the humanities. Any pitfalls which may lie ahead for the young scientist engaged in the most basic research—in the way of potential conflict with Marxist-Leninism—must seem a far-off danger, indeed, compared to immediate advantages.

One observer has characterized the resulting situation in the schools:[15]

Arithmetic, algebra, trigonometry, geometry, the laws of classical physics, chemistry—these remain the same whether politically biased phraseology is used or not. . . . Thus the teaching of these subjects suffers less than those fields where an interpretive bias can be freely applied. These conditions are but the starting point in a race in which the sciences win and the humanities lose in the Soviet educational setting.

This "other-worldliness" of science—particularly research—in which scientists of all ages find intellectual satisfaction by losing themselves completely in their work, tends to give the profession a unity and to diminish that conflict of generations more evident in other fields. Older scientists and professors are genuinely revered for knowledge and talents easily divorced from the particular politi-cal and social setting. They are people to be emulated, not displaced. So, in the case of young scientists, the rationalization of their men-tors' work under Stalin comes easily with the "proof" that their internationally recognized achievements were not directly tied to the revealed tyranny of the despot.

This is not to say, however, that the experimental faculty of scientific youth is uniformly walled off from Soviet life, or that the regime is content with the political attitudes of the scientific profes-sion. Young scientists appear to be some of the most caustic critics of the over-all system, if not of the place their own profession has in it. During a visit to Harvard University earlier this year, the young Soviet poet Evtushenko (an aggressive "reform" leader since 1956) said that he preferred to read his poetry to young scientists and engineers because of their "fresh minds." He believed his work enjoys its greatest success among this group, who read it in their spare time. And back in the troubled fall of 1956, the questioning of

Party truth seems to have been as widespread in the scientific faculties and institutes as in the liberal arts and humanities.

Mr. Khrushchev has thrown some light on this unrest among young scientists. In a recently published speech to "representatives of the Soviet intelligentsia," he singled out three anonymous young scientists to prove that no profession is too valuable to be above or beyond politics. He was discussing the disciplinary steps taken by the regime to still the storm of 1956. Mr. Khrushchev said the three renegades had been thrown out of the Party organization in their institute for "anti-Party" activities. When a "famous academician" had telephoned to plead for the future of these "talented boys," Mr. Khrushchev responded that their actions—unspecified—had not been children's play. He refused to relent, and went on to gloat over the calm such stern action had restored to the Soviet intelligentsia.[16]

Scientists, while perhaps more privileged than others of the intelligentsia, have much in common with certain young elite in other professions. The various frames of mind with which this young intelligentsia has emerged from the psychological upheaval of 1956 will be discussed in a section below.

The system of rewards through education has led to one of the greatest contradictions in Soviet society: although common physical labor is loudly praised, children of the powerful professional groups consider themselves failures if reduced to it. Their world is increasingly separated from that of the working class. Professional military officers vie with one another to get their sons into the elite Suvorov and Nakhimov military academies. The budding civilian aristocracy asserts its exclusiveness through unpublicized special secondary schools, like the one near Sokolniki Park in Moscow, where instruction is carried on exclusively in English, French, or German.[17] The ballet schools in Moscow and Leningrad have become another important status symbol. Generally speaking, the well-equipped ten-year secondary schools in the cities are far surer stepping stones to success than their seven-year counterparts still remaining in the countryside.

These educational gaps have conflicted sharply with the egalitarian features built into the over-all educational system from the revolutionary past. An upward mobility of workers and peasants has been encouraged in the name of that revolution. Everyone, regardless of social position or sex, has access to a free secondary-school education. Informal relationships between teacher and pupil, an overbearing emphasis in primary and secondary school on collectivism, and the enforcement of a common culture and language through centralized controls also have a leveling effect (although in the latter case poor instruction in Russian in the national republics hampers the advancement of these minorities in All-Union careers). This egalitarianism puts an unbearable strain on the higher educational establishments, which, by design of the economic plan, cannot absorb all secondary-school graduates.

With characteristic directness, Mr. Khrushchev has set out to undo Stalin's legacy through a wholesale shake-up of the educational system. "The chief and fundamental defect of our secondary and higher schools is the fact that they are detached from life," was his way of expressing it in September 1958.[18] He noted that in 1957 some 800,000 of 2,500,000 secondary-school graduates could not go on to higher education, but that they were prepared for nothing else. He discovered that only 30–40 percent of students in Moscow higher schools came from worker and peasant families. He found that many families had a "haughty and contemptuous" attitude toward physical work and that their children considered factory and farm work "beneath their dignity."

Mr. Khrushchev's proposed solution: "polytechnical" education in secondary schools to "establish ties with life" (work); a requirement that all students entering professional institutes and universities have at least two years' work experience; and a much closer alignment of specialized studies with practical work. Although this radical proposal was somewhat watered down through negotiation with Soviet educators, essential changes were introduced at the opening of the academic year 1959-1960.

There is no space here to analyze all the changes and their myriad exceptions. It is also too soon to judge the final impact of the reforms, for they will not be completed until 1964. It must be stressed, however, that their aim is *not* to reduce the number of students in higher education or to throw a huge supply of unskilled child labor onto the market. That intention might have been read into Mr. Khrushchev's first pronouncements. In practice, however, the reform (at least in the big cities) has added one year onto the old ten-year middle schools and shifted the curriculum for the last three years toward more "knowledge of production." In the school I visited in Moscow, this means that, beginning with the ninth grade, students spend four days a week in the school, and two days in a "patron factory" learning industrial skills. The principal, Maria Skyartsova, said the extra year had also allowed them to broaden

273

the teaching of literature, physics, chemistry, and mathematics, and to add a new course on world history since World War I.[19]

The aim is to change the orientation of students, not to reduce their numbers. Secondary-school graduates must be able to go into practical work; some may go on to full-time higher education in two years, others will have to be content with night school or vastly expanded correspondence courses.[20] The privileged ones who do become full-time day students are to be thoroughly reliable. According to new rules published early in 1959, they must have "good" recommendations from the Communist Party group, trade union, and Komsomol in the factory, as well as support from the factory director or collective farm board.[21]

Loopholes in the rules have already appeared. Given the power of the professional groups affected, those holes may be expected to become larger rather than smaller. First, at least 20 percent of first-year students are still to be chosen directly from secondary middle schools, although the new admission rules are notably ambiguous about students of the humanities.[22] This general 20 percent is a minimum figure; in fact, in 1959, 45 percent came directly from secondary schools.[23] This leeway in avoiding the work requirements obviously intensifies, rather than relieves, that cutthroat competition among well-placed parents which Mr. Khrushchev attacked in 1958. They still pull every possible string to get their children into the higher schools. Second, it is doubtful that those children are going to acquire that "worker's mentality" Mr. Khrushchev so admires by serving time in a factory or on a collective farm, or that the manual labor is going to aid their future careers. Sons and daughters of ensconced officialdom are already claiming confidentially that they will find suitable apprenticeships in good laboratories or executive offices. Some I met felt the elite would have no difficulty in getting the necessary recommendations for university entrance well ahead of the end of the two-year period.

The immediate political imperative for these reforms grew out of the unrest in 1956-1957. The debunking of Stalin, a near revolt in Poland, a real revolution in Hungary, armed intervention, and their aftermath produced a crisis of confidence among the young intelligentsia which clearly frightened Mr. Khrushchev and his hierarchy. That latent conflict of attitudes between generations had never before come so close to the surface, nor had it ever borne such ominous political overtones. Here were privileged students—particularly in the elite institutes and universities of Leningrad and

Moscow—who reacted to Khrushchev's revelations with a barrage of confused criticism, not with the embarrassed silence of their elders. They were young enough to question profoundly, old enough to know what to question. Although no one knows how many students displayed "unhealthy attitudes" in 1956-1957, enough existed at the very summit of Soviet education to warrant strong attacks in the press.[24] Soviet intervention in the Hungarian revolution and concomitant police coercion and social pressures at home finally silenced their protests but did not correct the condition underlying them. When Mr. Khrushchev later began demanding the resurrection of ties between the schools and "life," he really meant restoring firm ties between education and the Communist leadership.

Yet the reforms have come too late to affect the rebels of 1956-1957. Today they are no longer students, but they are not yet seasoned members of the "establishment." They are still grappling with the shock from the denunciation of Stalin, the thaw, the new freeze, and then the more careful relaxation since 1959. The result, I believe, is two recognizable strains in this part of the young intelligentsia: the "intellectuals" and the "men of action." Sometimes both strains coexist uneasily in the same person, sometimes not, but both types represent important ingredients in the present Soviet environment.

The "intellectuals" are the ones who have turned their backs on the establishment. They have not stopped questioning; they do not accept the materialistic red herrings Mr. Khrushchev has provided. They have a guilty conscience about the past. As one critic put it: "The intellectuals come straight out of Dostoevski, wailing and worrying about the future of society, about abstract ideas like 'justice,' but they can never take a decision, they can never do anything." In 1956, they were students who turned university seminars into serious political discussion groups and later expressed sympathy with the Hungarian rebels. Some were reprimanded, others were expelled, and still others, imprisoned. Now conditions have returned to "normal"; they are lonely, disillusioned and bitter.

"We have found out that when you beat your head against a stone wall, you break your head, not the wall," a dissident confided in Leningrad, more in sorrow than in anger. "The secret police have changed their tactics, they are more polite. Whether they call it arrest or education, however, their power amounts to the same thing. They are the stick, and the leadership of the Party wields it."

275

These young men and women in their middle and late twenties provide penetrating (and not always unsympathetic) insight into their more conformist fellows. "The bright young people see that this system has lasted over forty years and is growing steadily stronger," this ex-student said. "They believe the mistakes of the past are being corrected. So they conform. They fulfill all the outward norms, but in reality they are indifferent to politics. Do not think the lie they live is a conscious one. They honestly believe they are doing the right thing. Nagging doubts remain, but they are sub-conscious. That explains the loud enthusiasm for Khrushchev's great Seven Year Plan. If that succeeds, if the Soviet Union emerges with the highest standard of living in the world, these people feel their last troublesome doubts will die."

Some of the "intellectuals" are prone to analyze the effect of future prosperity. Some believe it will so mellow the "petty bour-geois" mentality of the ruling bureaucracy that nonconformists will be left more to their own devices. The progress of industrialization, the increased infiltration into Party ranks on the part of the rising technical intelligentsia, the slow expansion of "socialist legality" in day-to-day life—all these developments, they believe, may produce an "old-fashioned" dictatorship under which lip-service is still paid to totalitarian doctrine, but substantial areas of personal freedom exist. On the other hand, some of the more extreme believe the spiritual decay beneath the material prosperity will be the downfall of the whole Communist system.

"Have no illusions about the development of democracy here," were the bitter words of an anonymous man in his early thirties. We were sitting on high stools in the bar of the Sovietskaya Hotel, a meeting place in the capital of the "gilded youth." "The Russians are too submissive and disorderly. You speak of the power of edu-cation, but the men who rule us are educated. They are all philis-tines. Take the writers, for instance. No one compels them to write such trash now, but they are all afraid of losing their material well-being. I suppose when people live eight or ten to a room they have little time to think about anything besides finding some place of their own. But once they find that place and more besides, this sys-tem will just fall apart from moral decay." He paused to look around at the animated company on the dance floor from his perch on high, and added thoughtfully: "If Lenin returned, we would greet him in the way the Bible says Jesus was received: 'He came unto his own, but his own received him not.' We would say to him what Dostoevski's Grand Inquisitor told Christ: 'Go away and never come back!' "

Sometimes this lonely despair leads to a heart-rending outburst of patriotism. All else may fail them, but anguished love of Russia remains. "The troubles of today, yesterday, tomorrow—they are one small part of our history," suddenly exclaimed one troubled friend. "Mother Russia has survived before, it will survive again. Even if I were better off anywhere else in the world, I would still choose my Russia. I love it all—the crying, the laughing, drinking vodka, kissing one minute, cursing the next. You Westerners cannot understand that, can you?"

He did not wait for an answer. "Our temperament shocks you. We got it from the Mongols. Before the invasion, we were calm but disorganized. Today we are no longer calm, but we are still disorganized."

The "man of action" tends to dismiss these malcontents. For him, they are spiritual outcasts, powerless and meaningless. He compares them to Pasternak's Dr. Zhivago and himself to Lopatkin, the inventor-hero who triumphs over bureaucracy in Dudintsev's *Not By Bread Alone*. Zhivago is the man who claims to feel strongly about "right" and "wrong," but he irritates these young people because he can never choose sides. (Most have read only the selected excerpts of the book printed in the *Literaturnaya Gazeta* for 25 October 1958.) Lopatkin, on the other hand, is supposed to be "Soviet" to the core. He works and suffers through the system but comes out on the other side. One acquaintance compared Dudintsev's book to Chernyshevski's *What Is To Be Done?* of the last century. Both books had a "revolutionary" impact at different moments in history.

"Pasternak may write beautifully," he said, "but his ideas are outmoded. Dudintsev writes badly, but what he writes is social dynamite. I stayed up all night reading it."

The ideal young "man of action" claims he has reconciled himself to his system. He may be sensitive in his inner life, but he is utterly realistic about his social environment. Above all, he is diligent and ambitious. He respects power at home and abroad. By devious means he is working his way up in the ministries and industry to wield power. He joins the Communist Party to get ahead but thinks little of ideology. He reveres Lenin but admits that Marx could not foresee twentieth-century development. He (or she) aspires to dress like Americans, but neither envies nor fears the Americans. He wants to learn their technical—not their political—

skills. He wants peace, for the success of his economic gamble depends on it.

The cynicism about world revolution is barely disguised. One evening in late 1959 in the Praga Restaurant in Moscow, I commented on the number of Africans on the dance floor: "They must be among the revolutionaries being trained here."

"Too well dressed," my companion took up the jest.

"Part of the Khrushchev era," I ventured.

"Then they are certainly not for world revolution," he countered.

The Chinese offend this attitude, no matter how prized or necessary an ally they remain. They offend because they remind. There they are, suddenly aping all the early Soviet antics with a vengeance. They blaze with a genuine revolutionary fervor that most thoughtful Russians long ago abandoned to the slogan writers. They sound a clarion call from the past which many would like to ignore but cannot. China has become the revolutionary conscience, irritating because it can neither be stilled nor forgotten. Even those young Soviet zealots who yearn to return to the revolutionary ideals of Lenin are put off by the seemingly inhuman discipline of the Chinese. Their own deeply ingrained chauvinism is offended by the rising national power of 650 million Chinese to the East.

By 1959 at least some Soviet students no longer tried to put the gloss of solidarity over this distaste for the Chinese mentality. "They volunteer for the hardest work during vacations and then refuse payment," one student almost shouted over the clanking of metal in the self-service canteen of Moscow University. "They seem to love political meetings. Nothing official like ours. Someone simply decides a subject, they all excitedly agree, and then spend days priming for it."

A student in Leningrad related what had happened when his dormitory Komsomol organized a "voluntary Sunday." These days used to be popular during the 'twenties and 'thirties, when people donated leisure time to social work and slum clearance. Times have changed, he said.

"This project was rubbish clearance. Twenty out of twenty-two Chinese in the dormitory appeared. The other two were certified as ill. Do you know how many Russians turned up?" he asked rhetorically, and laughed. "Two out of three hundred—the organizer and the secretary of the Komsomol!"

The Komsomol or young Party workers are the basest offshoots of the "man of action" type. Superficially, they are sterling examples of the ideal "Soviet man." They are enrolled in the higher schools, receive diplomas in the various faculties, like all students, and at the same time lead all kinds of required "social work" on the side. They are the guardians of moral and political conformity in the Komsomol, to which more than 90 percent of students belong. In fact, this minority of activists are unbelievably cynical. One observer has set their number at between ten and fifteen percent of Soviet students.[25] Quite early in their student life, they sort out the mechanism of political power and fashion their careers accordingly. They learn the prescribed liturgy of Marxism-Leninism by heart and achieve progress through their ability to follow orders in enforcing it. Technically, they are elected to their Komsomol offices, but in fact, they are appointed by the bureaucratic apparatus. They are out of their depth when they have to engage in serious discussion outside the established framework, and they avoid it wherever possible. Mr. Khrushchev's sudden denunciation of Stalin and the moves to "democratize" Komsomol life in 1956 threw their ranks into complete disarray. The chain of command was temporarily broken, their confidence shattered, for no one from the top down knew any longer quite what line to enforce.

Now that the situation has been stabilized, these activists are firmly back in the saddle. They continue to exercise enormous power through the Komsomol role in the administration of the higher schools. The three-year work assignment a student receives on graduation depends as much on his Komsomol recommendation as on his grades. As noted above, the Komsomol also has new power in recommending which students are to be admitted to higher studies after working two years. The nature of these recommendations in turn depends on a student's willingness to serve in numerous "extra-curricular" activities such as the voluntary Sundays or summer work on collective farms. This power over a student's future naturally leads to Komsomol meddling in the most intimate human relationships. Activists extract public confessions, enforce punishment, make certain that no one publicly transgresses the bounds of propriety. A rebellious student risks expulsion from the Komsomol, and this is tantamount to social ostracism and expulsion from the university.

Mr. Khrushchev's great problem is how to foster the "spirit of revolution" among the educated young while maintaining this essential Komsomol control. The contradiction between means and

279

ends here sets the dimensions of potential tragedy in the Soviet leader's policy. On the one hand, he has fathered imaginative frontier plans for developing virgin lands in the Far East and South. Through these programs, more than half a million young people have moved out of Western Russia. He has also increased material incentives at all levels of society and has begun to fulfill some of the old promises about consumer prosperity. In the schools—beside polytechnical reforms for secondary day schools—he has instituted since 1956 a network of boarding schools (*internati*) which give increasing numbers of children from the age of seven on a proper "Communist upbringing" outside the family. By 1965 these *internati* are scheduled to have 2,500,000 out of the more than 30 million primary- and secondary-school pupils in the country. All these moves are attempts to maintain the momentum of Communist revolution without the excesses of Stalin.

On the other hand—working against those moves—Mr. Khrushchev's revelations about Stalin and the omnipresent hypocrisy of Soviet society today have destroyed much of the idealism necessary to maintain revolutionary momentum. Calculation rather than spontaneity is a young person's guide to success in the Soviet Union. Young people in both factory and university work out elaborate methods for escaping the snooping and the commands of the Komsomol without impairing their future. Among the students especially, service extorted in the name of socialist society is no longer an honor but a duty to be avoided wherever possible. Constant press attacks on the Komsomol's failure to enlist the "best" recruits for the virgin lands, coupled with reports of wild juvenile delinquency on the frontier, bear witness to the success of many evaders.[26] A growing emphasis on more persuasion and less coercion (with concomitant relaxation) only gives more scope for this evasion. The greater material rewards offered young people become incentives for more cynicism, not more idealism, because they can only be had by those who pay the price of enforced conformity.

Mr. Khrushchev's brand of Communism indisputably is opening up new vistas for Soviet youth. At the same time, the measure of those vistas must surely be the types of adults who emerge. The Soviet social scene is changing too rapidly to make any definite predictions. The answers lie well beyond Mr. Khrushchev, in what or who follows him. At this point, however, present signposts indicate, for better or for worse, that the on-coming adult generation little resembles the ideal Soviet Man of Communist mythology.

REFERENCES

1. For example, an article in *Komsomolskaya Pravda*, 23 June 1956, p. 4, lashes out at physical-education instructors who do anything to escape being sent out of Moscow to teach.

2. The emphasis here is on *ardent*, because mere membership of the Komsomol is little more than a formality today. By 1956 the Komsomol had 18,500,000 members aged fifteen to twenty-seven (A. N. Shelepin's "Report to the Twentieth Party Congress of the C.P.S.U.," *Pravda*, 22 February 1956, p. 8).

3. For instance, Mr. Khrushchev has given the best exposé of the snobbishness of the elite system in Soviet higher education: "Memorandum on School Reorganization," *Pravda*, 21 September 1958, pp. 2-3.

4. "On Slang and Fashionable Catchwords," *Neva*, Leningrad, September 1960, 9: 200-203, as translated in *Current Digest of the Soviet Press*, 1960, *XII, 46:* 15-16.

5. I saw these posters during a visit to Ivanovo at the end of July 1959.

6. "Patrol in Knee Pants," *Komsomolskaya Pravda*, 13 December 1960, p. 2.

7. *Ibid.*

8. "Soviet Jazz Awaits Its Composers," *Komsomolskaya Pravda*, 25 December 1960, p. 4.

9. "Patrol in Knee Pants," *op. cit.*

10. From the literary supplement of the Moscow University newspaper, *Raduga*, 1958.

11. See *Komsomolskaya Pravda*, 28 February 1957, p. 1, for a decree of the Central Committee of the Komsomol which comes close to admitting this conflict of generations. The younger generation has not been through "the severe school of revolutionary struggle," it says, and therefore takes the "great achievements of the Soviet people" for granted or does not appreciate them at all.

12. See George Z. F. Bereday, William W. Brickman, Gerald H. Read (editors), *The Changing Soviet School* (Cambridge: Riverside Press, 1960), p. 423, for the impressions of a group of American educators as to the effect of disorganized family life on delinquency in the Soviet Union.

13. See George Z. F. Bereday and Jann Pennar, *The Politics of Soviet Education* (New York: Frederick A. Praeger, 1960), ch. 4, "Class Tensions in Soviet Education," pp. 57-88, for a more comprehensive analysis of the conflict between egalitarianism and status in Soviet education.

14. Nicholas DeWitt, *Soviet Professional Manpower* (Washington: National Science Foundation, 1955), pp. 167-169, 217. Mr. DeWitt also found that, while Soviet higher education graduated 40 percent less than American

colleges and universities, only 25 percent of American bachelor degrees went to scientific and engineering students.

15. *Ibid.*, p. 41.

16. Although this speech was actually delivered 17 July 1960, it was not published until May 1961. N. S. Khrushchev, "Toward New Success of Literature and Art," *Kommunist*, May 1961, 7: 6-7.

17. *Ibid.*, p. 120. David Burg, a student in Moscow until 1956, writes that the children of Mr. Malenkov attended the English-language school near Sokolniki Park while he was premier. This confirms my personal observations during 1959 about the exclusiveness of this school.

18. Mr. Khrushchev's "Memorandum on School Reorganization," *op. cit.*

19. Middle School No. 49, which I visited on 28 September 1959.

20. P. I. Polukhin, a spokesman for the All-Union Ministry of Higher Education, told me on 4 December 1959 that one-half of the two million students then enrolled in higher education were in night school or taking correspondence courses.

21. *Pravda,* 4 April 1959, p. 4.

22. *Ibid.:* this article seems to suggest that *all* students in these disciplines must have the preliminary work experience.

23. According to P. I. Polukhin in the interview noted above.

24. David Burg, "Observations on Soviet University Students," *Dædalus,* Summer 1960, pp. 520-540. This article gives a detailed report on student reactions to the events of 1956. See also S. V. and P. Utechin, "Patterns of Nonconformity," *Problems of Communism*, 1957, 3: 15-23. *Komsomolskaya Pravda,* 16 December 1956 and 28 December 1956, tells of handwritten magazines in Leningrad bearing such names as *Fresh Voices* and *Heresy.* See also "Strength and Faith," *Izvestia,* 6 September 1959, p. 4, for an account of the Soviet handling of another student conspiracy.

25. David Burg, *op. cit.*, p. 525.

26. See *Literatura i Zhizn,* Moscow, 17 February 1961, as reported in *The New York Times,* 18 February 1961, p. 1, for an account of gang warfare and murder among youths in Vladivostok.

Notes on Contributors

BRUNO BETTELHEIM, born in Vienna in 1903, is professor of educational psychology and principal of the Orthogenic School at the University of Chicago. Since coming to the United States in 1939 he has published a number of studies in the psychoanalytic field, especially on child psychology: *The Dynamics of Prejudice* (with Morris Janowitz); *Love is Not Enough—The Treatment of Emotionally Disturbed Children; Symbolic Wounds; Truants from Life;* and most recently, *The Informed Heart.*

ROBERT COLES is a physician. Born in Boston in 1929, he completed his psychiatric training at Harvard in 1961, after serving for two years at an Air Force hospital in Mississippi. What he had seen there in the communities surrounding the air base induced him to move to Georgia in order to study how "Negro and white people, and their children, get along with themselves and with one another in a new South."

REUEL DENNEY, born in New York City in 1913, professor of social sciences at the University of Chicago, is currently at the East-West Center of the University of Hawaii. As a poet, a critic, and a chronicler, his wide-ranging activities in the humanities parallel his concern with popular culture, the mass audience, and the ways in which taste is formed. His book, *The Astonished Muse*, deals with the mass media and the individual choices and identifications these promote. With David Riesman and Nathan Glazer, he is co-author of *The Lonely Crowd.*

S. N. EISENSTADT, born in Warsaw in 1923, is professor of sociology at the Hebrew University in Jerusalem. He has also taught at the University of Chicago and the University of Oslo. His concern with the problems of immigration, youth and youth culture, and political modernization are the subjects of his books: *Absorption of Immigrants; From Generation to Generation; Bureaucracy and Bureaucratization; Essays on Sociological Aspects of Political and Economic Development; The Social and Political Systems of Centralized Empires* (forthcoming).

ERIK H. ERIKSON, born in Frankfurt-am-Main in 1902, of Danish parentage, is professor of human development and lecturer on psychiatry at Harvard University. Among his publications are: *Childhood and Society; Young Man Luther; Studies in the Interpretation of Play; The Yurok; Childhood and World-Image;* and *Identity and the Life Cycle.* The studies of which the present chapter forms a part have been supported by a grant from the Ford Foundation to the Austen Riggs Center.

THE HON. ARTHUR GOLDBERG is Associate Justice of the United States Supreme Court. He wrote this article while he was United States Secretary of Labor. Born in Chicago in 1908, he graduated from the Northwestern University Law School in 1929 and has since had a distinguished career as a lawyer and a counsel to labor organizations.

JOSEPH F. KAUFFMAN is Director of Training for the Peace Corps. Born in Providence, R. I., in 1921, he holds a doctorate from Boston University. He has served as Dean of Students and Assistant to the President at Brandeis

University and as Executive Vice President of the Jewish Theological Seminary of America.

KENNETH KENISTON, born in Chicago in 1930, is lecturer in social relations at Harvard University. From 1951 to 1953 he was a Rhodes Scholar. He has contributed articles in the field of clinical research in psychology and the psychological effects of cultural and historical change to *The American Scholar,* the *Journal of Social and Abnormal Psychology,* and *Commentary.*

ROBERT JAY LIFTON, born in New York City in 1926, has recently been appointed to the chair of the Foundations' Fund for Research in Psychiatry at Yale University. Currently he is conducting a two-year psychological study of Japanese youth at Tokyo and Kyoto, in which he concerns himself with the relation between individual psychology and cultural-historical change. His recent book, *Thought Reform and the Psychology of Totalism,* based on research in Hong Kong in 1954-1955, deals with Chinese Communist "brainwashing" and related psychological issues.

KASPAR D. NAEGELE, born in Stuttgart in 1923, is professor of sociology at the University of British Columbia. He has also taught at Harvard University and the University of Oslo. His publications include: *Canadian Society* (with others); and *Theories of Society* (with Talcott Parsons, Edward Shils, and Jesse R. Pitts).

TALCOTT PARSONS, born in Colorado Springs in 1902, is professor of sociology at Harvard University, and has taught intermittently at other institutions, including Cambridge University. His particular concerns are with the sociology of religion and of law, problems of aging, and political sociology. Besides his numerous contributions to periodical literature, his most recent books are: *Toward a General Theory of Action; The Social System; Essays in Sociological Theory;* and *Structure and Process in Modern Societies.*

GEORGE SHERMAN, born in Boston in 1930, is currently covering Latin-American affairs for the Washington *Evening Star.* From 1956 to 1960 he was traveling correspondent in East Europe and the Soviet Union for the London *Observer.* A graduate of Dartmouth College, he has a master's degree from the School of International Affairs and the Russian Institute, Columbia University, and a graduate degree from St. Antony's College at Oxford University. He has contributed articles on Soviet and East European politics to various English and American periodicals.

LAURENCE WYLIE, born in Indianapolis in 1909, is C. Douglas Dillon Professor of the Civilization of France at Harvard University. He has taught at Haverford College and other institutions, and from 1955 to 1956 he was a Ford Faculty Fellow; from 1957 to 1958, a Guggenheim Fellow. His special concerns are with community studies in France and with French literature as the vehicle of French culture. His books are: *Saint-Marc Girardin Bourgeois;* and *Village in the Vaucluse.*